Linear
Programming
with BASIC
and FORTRAN

Linear Programming with BASIC and FORTRAN

CARVEL S. WOLFE

U.S. Naval Academy

Reston Publishing Company, Inc.
A Prentice-Hall Company
Reston, Virginia

Library of Congress Cataloging in Publication Data

Wolfe, Carvel S.
   Linear programming with BASIC and FORTRAN.

   1. Linear programming. 2. Basic (Computer program
language) 3. FORTRAN (Computer program language)
I. Title.
T57.74.W64 1985   519.7'2   84-24758
ISBN 0-8359-4082-9

IBM ® is a registered trademark of International Business Machines Corporation.

Editorial production supervision and interior design by Carolyn Ormes

© 1973, 1985 by Reston Publishing Company, Inc.
A Prentice-Hall Company
Reston, Virginia 22090

10 9 8 7 6 5 4 3 2 1

PRINTED IN THE UNITED STATES OF AMERICA

Contents

Preface

This book is written as an introduction to the important topics of linear and integer programming. Readers in the fields of operations analysis, computer science, analytical management, economics, game theory, government, and other related areas find that today their large problems are solved on electronic digital computers. Many of these problems use the elegant techniques of linear programming. Integer programming, in turn, relies on the foundation of linear programming.

A primary objective of this book is to make these techniques clear and usable to the reader. A part of the beauty of linear programming is its simplicity. Only elementary mathematics is needed, algebra and geometry. The Simplex Method of George Dantzig is the key idea involved in solving these problems. All of the computations in the Simplex Method use simple formulas, and the process can be easily programmed on a computer. The ideas are developed in this text and are simultaneously illustrated with examples. Then the steps are formulated into algorithms, which are programmed for running on a computer. One of the goals has been to develop a convenient notation that facilitates the computer programming and makes the answers readable at a glance.

All of this material has been used and refined for a number of years in the linear programming courses at the U.S. Naval Academy. Students in these courses rapidly learn to run their own programs on a high speed computer and to solve various practical problems. They use a time sharing system of hundreds of terminals connected to a high speed Honeywell dual processor. However, the programs can be run on any modern computer, including the many personal computers now available.

The problem exercises are an integral part of the text. There are a number of short problems that can be done with only pencil and paper while one is learning technique. The considerable variety of problems will create interest and challenge ingenuity. Some of these exercises extend the material in the text and offer new ideas. The text can thus adapt to many levels of preparation. The longer problems require a digital computer. To aid in the computer programming, FORTRAN and BASIC subroutines for the necessary calculations are presented and explained in Chapter 5.

A tried and recommended approach is to combine the study of the Simplex Method in Chapter 4 with the writing of computer programs to do the computations, as explained in Chapter 5. Then, as the algorithms are developed you can enjoy seeing the output on your computer. Subroutines used in the Primal-Dual Method of Chapter 8 are presented in the Appendix. It is thus recommended that you write your own main programs to enter data, call the subroutines needed, and print results on your machine. Programs to carry out each algorithm are available from the publisher on a disk that runs on an IBM® PC, or compatible computer.

A special effort has been made to develop and simplify the algorithms of integer programming. Integer programming is technically the most difficult area of mathematical programming because of problems of convergence, round-off error, and search time. All of the current methods have their short-comings, as explained in Chapter 11. Programs in BASIC for using fractional cutting planes and for using branch and bound to get integer answers are also on the available disk.

The present text is expanded and updated from an earlier version, which appeared in 1973 and was entitled, "Linear Programming with FORTRAN."

I give many thanks to my colleague W. Charles Mylander, who read and helped to edit the new material. Also, I thank my wife, Margaret, who did most of the proofreading.

C.S.W.

CHAPTER 1

Linear Algebra

Many of the topics of linear algebra are familiar from elementary mathematics, where small sets of linear equations were solved by methods such as graphing and elimination. The ideas of vectors and their combinations were handled both by geometry and by algebra. A formal course in linear algebra often tries to organize these concepts in an abstract algebraic structure.

This chapter summarizes the linear algebra operations that are a useful background for linear programming without most of the abstraction. Matrices are introduced as representations of linear transformations, and matrix operations are then used to solve systems of linear equations.

1.1 VECTORS AND VECTOR SPACES

Vectors are familiar to those using such physical quantities as force, velocity, and acceleration. These quantities have both magnitude and direction. It is just as important to know in what direction a force acts as to know its size. In two or three dimensions the algebra of vectors may be understood by drawing pictures, such as the triangle, for adding two vectors. In generalizing to an arbitrary number of dimensions, it is more convenient to define a vector in terms of its components. If we think of a vector as a directed line segment emanating from the origin of a rectangular coordinate system, then its *components* are the projections of the vector on each of the coordinate axes. These projections form an ordered *n*-tuple of real numbers that may be considered coordinates of the head of the vector. For any positive integer *n* the collection of all possible *n*-tuples of real numbers forms an *n*-dimensional geometry called real *n*-space, which will be denoted by R^n.

DEFINITION 1.   For any positive integer *n*, a *vector*, **X**, in space R^n is an ordered *n*-tuple of real numbers.

$$\mathbf{X} = (x_1, x_2, \ldots, x_n)$$

An example of a vector from the world of business is found in the production of parts. Suppose a production line makes 37 different parts and the foreman keeps track of the number of each of the 37 parts that are produced in one day. Then the daily output from this line is a vector with 37 components. Each component is one of the parts and its value on a given day is the number of that part produced.

Vector addition is defined as follows:

DEFINITION 2.   The *sum* of any two vectors **X** and **Y** in R^n is the vector in R^n whose components are the sums of the corresponding components of **X** and **Y**.

$$\mathbf{X} + \mathbf{Y} = (x_1, x_2, \ldots, x_n) + (y_1, y_2, \ldots, y_n) = (x_1 + y_1, x_2 + y_2, \ldots, x_n + y_n)$$

In the example of production of parts, let **X** be the output for one day and **Y** be the output for another day. Then the sum **X** + **Y** is the total production of all 37 parts over both days.

A vector may be multiplied by any real number according to the next definition.

DEFINITION 3.   *Multiplication by a scalar* is defined for all vectors **X** in R^n and all real numbers α by the vector in R^n whose components are α times the components of **X**.

$$\alpha \mathbf{X} = \alpha(x_1, x_2, \ldots, x_n) = (\alpha x_1, \alpha x_2, \ldots, \alpha x_n)$$

A number of properties follow immediately from these definitions. First, vector addition is commutative, that is,

$$\mathbf{X} + \mathbf{Y} = \mathbf{Y} + \mathbf{X}$$

The commutativity follows since the sums of real numbers in each component are commutative. Vector addition is also associative, that is,

$$\mathbf{X} + (\mathbf{Y} + \mathbf{Z}) = (\mathbf{X} + \mathbf{Y}) + \mathbf{Z}.$$

Again the truth follows from the corresponding property of real numbers in each component. Multiplication by a scalar is easily shown to be distributive by using definitions 2 and 3.

$$\alpha(\ \mathbf{X} + \mathbf{Y}\ ) = [\ \alpha(\ x_1 + y_1\ ),\ \alpha(\ x_2 + y_2\ ),\ \ldots,\ \alpha(\ x_n + y_n\ )\ ]$$
$$= (\ \alpha x_1 + \alpha y_1,\ \alpha x_2 + \alpha y_2,\ \ldots,\ \alpha x_n + \alpha y_n\ )$$
$$= (\ \alpha x_1,\ \alpha x_2,\ \ldots,\ \alpha x_n\ ) + (\ \alpha y_1,\ \alpha y_2,\ \ldots,\ \alpha y_n\ )$$
$$= \alpha\mathbf{X} + \alpha\mathbf{Y}$$

By a similar proof, for all real numbers α and β

$$(\ \alpha + \beta\ )\mathbf{X} = \alpha\mathbf{X} + \beta\mathbf{X}.$$

It is clear from the definition of multiplication by a scalar that $1 \cdot \mathbf{X} = \mathbf{X}$. We shall define a *zero vector*, $\mathbf{0}$, by

$$\mathbf{X} + \mathbf{0} = \mathbf{0} + \mathbf{X} = \mathbf{X}$$

for all vectors $\mathbf{X}$ in R^n. Geometrically the zero vector may be considered as a point at the origin such that all of the components of this vector are zero.

$$\mathbf{0} = (\ 0, 0, \ldots, 0\ )$$

From the distributive laws we get the relations between the zero of the real number field and the zero vector.

$$\alpha\mathbf{X} + \alpha\mathbf{0} = \alpha(\ \mathbf{X} + \mathbf{0}\ ) = \alpha\mathbf{X} = \alpha\mathbf{X} + \mathbf{0}$$

Canceling $\alpha\mathbf{X}$ from both sides leaves $\alpha\mathbf{0} = \mathbf{0}$ for all real α. Thus any multiple of the zero vector is the zero vector. Similarly

$$\alpha\mathbf{X} + 0\mathbf{X} = (\ \alpha + 0\ )\mathbf{X} = \alpha\mathbf{X} = \alpha\mathbf{X} + \mathbf{0}.$$

Again canceling $\alpha\mathbf{X}$ leaves the result $0\mathbf{X} = \mathbf{0}$ for all $\mathbf{X}$. This means the real number zero times any vector is the zero vector. Finally the additive inverse for any vector $\mathbf{X}$ is $(-1)\mathbf{X}$ or $-\mathbf{X}$.

$$\mathbf{X} + (-1)\mathbf{X} = 1 \cdot \mathbf{X} + (-1)\mathbf{X} = [\ 1 + (-1)\ ]\mathbf{X} = 0\mathbf{X} = \mathbf{0}$$

Every vector added to its inverse gives the zero vector, sometimes called the *null* vector.

If we collect all of these ideas together we can formally define a vector space or linear space. The scalar numbers used are called a *field*, *F*. For our purposes the field, *F*, will be the real number system including all the rational and irrational numbers.

DEFINITION 4. A *vector space*, **V**, over a field, *F*, is a set of elements (vectors) that satisfy the following properties:

1. For every **X** and every **Y** in **V**, there is a unique vector **X** + **Y** in **V** called the sum.
2. For every **X**, **Y**, and **Z** in **V**, the associative law holds, thus **X** + (**Y** + **Z**) = (**X** + **Y**) + **Z**.
3. There is an identity element in **V** called **0** such that **X** + **0** = **0** + **X** = **X** for all **X** in **V**.
4. There is an inverse element in **V** called − **X** such that **X** + (−**X**) = **0** for all **X** in **V**.
5. The commutative law holds, **X** + **Y** = **Y** + **X** for all **X** and **Y** in **V**.
6. For every **X** in **V** and every α in field *F*, there is a unique vector α**X** in **V** called the product of α and **X**.
7. The distributive laws hold, α(**X** + **Y**) = α**X** + α**Y** and (α + β)**X** = α**X** + β**X** for all **X** and **Y** in **V**, all α and β in field *F*.
8. $1 \cdot$ **X** = **X** for all **X** in **V**.
9. ($\alpha\beta$)**X** = α(β**X**) for all **X** in **V** and all α and β in the field *F*.

In the terminology of algebra the first five axioms of a vector space say that the vectors form a *commutative group*. This group is closed under multiplication by field elements according to axiom 6. Axiom 7 says multiplication by a field element is distributive over vector sums and multiplication by a vector is distributive over the sum of field elements. Axiom 9 is an associative law for the product of a vector with a pair of field elements.

It has already been shown that *n*-tuples of real numbers satisfy all nine axioms over the real number field, so R^n is a vector space for each positive integer *n*. There are many other examples of vector spaces. The set of all real valued functions over any given domain is a vector space where the field in question is the real numbers and the sum of functions is defined to be the sum of their functional values. Vector spaces may use the complex numbers for their scalar field. Modern mathematics and physics frequently have for their foundation some appropriate vector space.

DEFINITION 5. A *subspace* **S** of a vector space **V** is a subset of the elements of **V** that satisfies in itself all nine axioms of a vector space with respect to the operations in **V**.

Actually in a subset **S** of **V**, it is only necessary to check the closure of **S** under addition and the closure of **S** under multiplication by a scalar to show that **S** is a subspace of **V**. The rest of the axioms are true since the elements of **S** are elements of **V**. Each subspace must contain the origin. Thus the proper subspaces of R^3 are any plane through the origin, any line through the origin, or the origin itself known as the *null subspace*. R^n is also considered a subspace of itself.

1.2   BASIS FOR A VECTOR SPACE

This section discusses the structure of a vector space and its subspaces in terms of a set of vectors called a *basis*. First we will define some preliminary concepts.

DEFINITION 6.   A vector $\mathbf{Y}$ in R^n is a *linear combination* of the vectors $\mathbf{X}_1, \mathbf{X}_2, \ldots, \mathbf{X}_k$ if it can be written as

$$\mathbf{Y} = c_1\mathbf{X}_1 + c_2\mathbf{X}_2 + \cdots + c_k\mathbf{X}_k$$

where the c's are k real numbers.

As an example let $\mathbf{Y} = (2, 3, -1)$, $\mathbf{X}_1 = (1, 2, 3)$ and $\mathbf{X}_2 = (1, 3, 10)$. The constants must satisfy

$$2 = c_1 + c_2$$
$$3 = 2c_1 + 3c_2$$
$$-1 = 3c_1 + 10c_2$$

which has solution $c_1 = 3$ and $c_2 = -1$. Thus

$$\mathbf{Y} = 3\mathbf{X}_1 - \mathbf{X}_2$$

satisfies definition 6.

DEFINITION 7.   A subspace $\mathbf{V}$ of R^n is *spanned* by the set of vectors $\{\mathbf{X}_1, \mathbf{X}_2, \ldots, \mathbf{X}_k\}$ in $\mathbf{V}$, provided every vector in $\mathbf{V}$ is a linear combination of $\mathbf{X}_1, \mathbf{X}_2, \ldots, \mathbf{X}_k$.

An example of a spanning set for R^3 is $\mathbf{X}_1 = (0, 1, 1)$, $\mathbf{X}_2 = (1, 0, 1)$, $\mathbf{X}_3 = (1, 1, 0)$. To show that $\{\mathbf{X}_1, \mathbf{X}_2, \mathbf{X}_3\}$ spans R^3, let $\mathbf{Y} = (y_1, y_2, y_3)$ be any vector in R^3 and find constants such that

$$\mathbf{Y} = c_1\mathbf{X}_1 + c_2\mathbf{X}_2 + c_3\mathbf{X}_3.$$

In component form this vector equation is equivalent to the three equations

$$y_1 = 0c_1 + 1c_2 + 1c_3$$
$$y_2 = 1c_1 + 0c_2 + 1c_3$$
$$y_3 = 1c_1 + 1c_2 + 0c_3.$$

A solution to this system for any y_1, y_2, and y_3 is

$$c_1 = \tfrac{1}{2}(-y_1 + y_2 + y_3)$$
$$c_2 = \tfrac{1}{2}(y_1 - y_2 + y_3)$$
$$c_3 = \tfrac{1}{2}(y_1 + y_2 - y_3).$$

For the vector $\mathbf{Y} = (1, 2, 3)$ the result gives

$\mathbf{Y} = 2\mathbf{X}_1 + \mathbf{X}_2 + 0\mathbf{X}_3.$

If the component equations do not have a solution for some choice of $\mathbf{Y}$, then the given set of vectors does not span the space.

DEFINITION 8.   A set of distinct vectors $\mathbf{S} = \{\ \mathbf{X}_1,\ \mathbf{X}_2,\ ...,\ \mathbf{X}_k\ \}$ in subspace $\mathbf{V}$ of R^n is *linearly dependent* if k real constants not all zero can be found such that

$$c_1\mathbf{X}_1 + c_2\mathbf{X}_2 + ... + c_k\mathbf{X}_k = \mathbf{0}. \qquad (1)$$

If equation (1) can only be satisfied with all k constants equal to zero, then the set $\mathbf{S}$ is *linearly independent.* To show whether a given set of vectors is linearly independent or linearly dependent, we have to prove, in the independent case, that all of the constants in (1) must be zero or, in the dependent case, that some constant in (1) is different from zero.

Consider the following two sets of vectors in space R^3, the first of which is linearly independent and the second of which is linearly dependent:

$\mathbf{S}_1 = \{\ \mathbf{X}_1,\ \mathbf{X}_2\ \}$ where $\mathbf{X}_1 = (\ 1, 2, 3\ )$ and $\mathbf{X}_2 = (\ 3, 2, 1\ )$

$\mathbf{S}_2 = \{\ \mathbf{X}_1,\ \mathbf{X}_2,\ \mathbf{X}_3\ \}$ where $\mathbf{X}_1 = (\ 1, 1, 2\ )$, $\mathbf{X}_2 = (\ 3, 0, -1\ )$,

$$\text{and } \mathbf{X}_3 = (\ -3, 3, 8\ ).$$

To show that $\mathbf{S}_1$ is linearly independent, we must prove that $c_1 = c_2 = 0$ in

$c_1\mathbf{X}_1 + c_2\mathbf{X}_2 = \mathbf{0}.$

In component form

$1c_1 + 3c_2 = 0$

$2c_1 + 2c_2 = 0$

$3c_1 + 1c_2 = 0.$

The unique solution to this system is $c_1 = c_2 = 0$.

On the other hand, to show that $\mathbf{S}_2$ is linearly dependent, we must find a set of constants not all zero such that

$c_1\mathbf{X}_1 + c_2\mathbf{X}_2 + c_3\mathbf{X}_3 = \mathbf{0}.$

In component form

$1c_1 + 3c_2 - 3c_3 = 0$

$1c_1 + 0c_2 + 3c_3 = 0$

$2c_1 - 1c_2 + 8c_3 = 0.$

A solution to this system is $c_1 = -3$, $c_2 = 2$, $c_3 = 1$, so that the sum

$$-3\mathbf{X}_1 + 2\mathbf{X}_2 + \mathbf{X}_3 = \mathbf{0}. \tag{2}$$

In addition to using definition 8, there is another way to test for linear dependence. Notice from equation (2) that $\mathbf{X}_3$ may be expressed as a linear combination of $\mathbf{X}_1$ and $\mathbf{X}_2$, namely

$$\mathbf{X}_3 = 3\mathbf{X}_1 - 2\mathbf{X}_2. \tag{3}$$

Whenever equation (1) is true with nonzero constants, any one of the vectors with a nonzero coefficient may be written as a linear combination of the remaining vectors. The converse to this statement is also true, since any linear combination such as (3) can be rewritten in form (2). Thus a test for linear dependence of a set $\mathbf{S}$ is whether one vector in $\mathbf{S}$ may be written as a linear combination of the remaining vectors in $\mathbf{S}$.

DEFINITION 9. A set of vectors $\mathbf{S} = \{ \mathbf{X}_1, \mathbf{X}_2, \ldots, \mathbf{X}_k \}$ in any subspace $\mathbf{V}$ of R^n is a *basis* for $\mathbf{V}$ if $\mathbf{S}$ both spans $\mathbf{V}$ and is linearly independent.

An example of a basis in space R^3 is the set $\{ (0, 1, 1), (1, 0, 1), (1, 1, 0) \}$. It has already been shown that this set of vectors spans R^3, and it is easy to verify by definition 8 that this set is also linearly independent.

Because of its simplicity the following basis in R^3 is known as the *standard basis*.

$$\mathbf{S} = \{ (1, 0, 0), (0, 1, 0), (0, 0, 1) \}$$

where the *unit vectors*, or vectors of length one, are called $\mathbf{i} = (1, 0, 0)$, $\mathbf{j} = (0, 1, 0)$, and $\mathbf{k} = (0, 0, 1)$. Bases have the following properties that can be proved.

1. If $\mathbf{S} = \{ \mathbf{X}_1, \mathbf{X}_2, \ldots, \mathbf{X}_k \}$ is a basis for subspace $\mathbf{V}$ of R^n, then every vector $\mathbf{Y}$ in $\mathbf{V}$ can be written uniquely as a linear combination of the vectors in $\mathbf{S}$.
2. Two bases for the same subspace must have the same number of vectors. Since the number of vectors in any basis for subspace $\mathbf{V}$ is unique, this number is called the *dimension* of $\mathbf{V}$.
3. If $\mathbf{V}$ is a subspace of dimension k, then any set of k linearly independent vectors in $\mathbf{V}$ is a basis for $\mathbf{V}$.
4. If $\mathbf{V}$ is a subspace of dimension k then any set of k vectors that span $\mathbf{V}$ is a basis for $\mathbf{V}$.

By property 1 any vector $\mathbf{Y}$ in $\mathbf{V}$ may be written

$$\mathbf{Y} = c_1\mathbf{X}_1 + c_2\mathbf{X}_2 + \cdots + c_k\mathbf{X}_k$$

where the constants are uniquely determined. These constants (c_1, c_2, ..., c_k) are known as the *coordinates* of **Y** with respect to the given basis. The coordinates of **Y** with respect to the standard basis in R^n are the same as the components of **Y**.

1.3 LINEAR TRANSFORMATIONS

Matrices and the way they combine arise in a natural way in the study of linear transformations from one vector space to another. A transformation or mapping is called *linear* in going from the first to the second vector space if it preserves vector sums and multiplication of a vector by a scalar. This means that the mapping of a sum is the sum of the mappings, and the mapping of a product by a scalar is the scalar times the mapping.

DEFINITION 10.   A *linear transformation* T from R^n to R^m is a function that associates to each vector in R^n a corresponding unique vector in R^m and satisfies the two rules:

$$T(\alpha \mathbf{X}) = \alpha T(\mathbf{X}) \tag{1}$$

$$T(\mathbf{X+Y}) = T(\mathbf{X}) + T(\mathbf{Y}). \tag{2}$$

An example of a linear transformation from R^3 to R^2 is

$$T(x_1, x_2, x_3) = (x_1 + 2x_2, 2x_3 - 3x_2).$$

This function tells how to compute the two coordinates in R^2 from the three given coordinates in R^3. It is shown to be linear using the distributive, associative, and commutative laws, as follows.

$$
\begin{aligned}
T(\alpha \mathbf{X}) &= T(\alpha x_1, \alpha x_2, \alpha x_3) \\
&= (\alpha x_1 + 2\alpha x_2, 2\alpha x_3 - 3\alpha x_2) \\
&= \alpha(x_1 + 2x_2, 2x_3 - 3x_2) \\
&= \alpha T(\mathbf{X}) \\
T(\mathbf{X+Y}) &= T(x_1 + y_1, x_2 + y_2, x_3 + y_3) \\
&= [(x_1 + y_1) + 2(x_2 + y_2), 2(x_3 + y_3) - 3(x_2 + y_2)] \\
&= (x_1 + y_1 + 2x_2 + 2y_2, 2x_3 + 2y_3 - 3x_2 - 3y_2) \\
&= (x_1 + 2x_2, 2x_3 - 3x_2) + (y_1 + 2y_2, 2y_3 - 3y_2) \\
&= T(\mathbf{X}) + T(\mathbf{Y})
\end{aligned}
$$

Another example that is easily shown to be linear is a projection of vectors in R^3 onto the xy plane.

$$T(x_1, x_2, x_3) = (x_1, x_2)$$

A transformation using the same spaces that is not linear is

$$T(\, x_1, x_2, x_3\,) = (\, 2x_1 + 3, x_2\,).$$

Rule (1) is violated.

$$
\begin{aligned}
T(\, \alpha\mathbf{X}\,) &= T(\, \alpha x_1, \alpha x_2, \alpha x_3\,)\\
&= (\, 2\alpha x_1 + 3, \alpha x_2\,)\\
&\neq \alpha T(\, \mathbf{X}\,)\\
&= \alpha(\, 2x_1 + 3, x_2\,)\\
&= (\, 2\alpha x_1 + 3\alpha, \alpha x_2\,)
\end{aligned}
$$

Rule (2) is also violated.

$$
\begin{aligned}
T(\, \mathbf{X+Y}\,) &= T(\, x_1 + y_1, x_2 + y_2, x_3 + y_3\,)\\
&= [\, 2(\, x_1 + y_1\,) + 3, (x_2 + y_2)\,)\,]\\
&= (\, 2x_1 + 2y_1 + 3, x_2 + y_2\,)\\
&\neq T(\, \mathbf{X}\,) + T(\, \mathbf{Y}\,)\\
&= (\, 2x_1 + 3, x_2\,) + (\, 2y_1 + 3, y_2\,)\\
&= (\, 2x_1 + 2y_1 + 6, x_2 + y_2\,)
\end{aligned}
$$

DEFINITION 11. A real *matrix* is a rectangular array of real numbers.

$$
\begin{pmatrix}
a_{11} & a_{12} & \cdots & a_{1n}\\
a_{21} & a_{22} & \cdots & a_{2n}\\
\cdot & \cdot & & \cdot\\
\cdot & \cdot & & \cdot\\
\cdot & \cdot & & \cdot\\
a_{m1} & a_{m2} & \cdots & a_{mn}
\end{pmatrix} = (\, a_{ij}\,)
$$

The first subscript i is called the row index and it indicates in which row a number is found. The second subscript j, known as the column index, tells in which column a number is found. Subscript i runs from 1 to m and subscript j from 1 to n where we have m rows and n columns. Each row may be thought of as a row vector in R^n and each column may be considered a column vector in R^m.

To see how rectangular arrays occur in linear transformations, consider a mapping $T(\, \mathbf{X}\,) = \mathbf{Y}$ from R^3 to R^3. By property 1 in section 1.2 a vector $\mathbf{X}$ in R^3 may be expressed as a linear combination of basic vectors in R^3. In what follows any basis may be used; however, for convenience, the standard basis will be assumed with unit vectors

$$\mathbf{i} = (\ 1, 0, 0 \ )$$
$$\mathbf{j} = (\ 0, 1, 0 \ )$$
$$\mathbf{k} = (\ 0, 0, 1 \ ).$$

Then

$$
\begin{aligned}
T(\ \mathbf{X} \ ) &= T(\ x_1, x_2, x_3 \ ) \\
&= T(\ x_1\mathbf{i} + x_2\mathbf{j} + x_3\mathbf{k} \ ) \\
&= x_1 T(\ \mathbf{i} \ ) + x_2 T(\ \mathbf{j} \ ) + x_3 T(\ \mathbf{k} \ ).
\end{aligned}
$$

Suppose

$$T(\ \mathbf{i} \ ) = a_{11}\mathbf{i} + a_{21}\mathbf{j} + a_{31}\mathbf{k}$$
$$T(\ \mathbf{j} \ ) = a_{12}\mathbf{i} + a_{22}\mathbf{j} + a_{32}\mathbf{k}$$
$$T(\ \mathbf{k} \ ) = a_{13}\mathbf{i} + a_{23}\mathbf{j} + a_{33}\mathbf{k}.$$

Combining and collecting terms

$$
\begin{aligned}
T(\ \mathbf{X} \ ) &= (\ a_{11}x_1 + a_{12}x_2 + a_{13}x_3 \ )\mathbf{i} \\
&\quad + (\ a_{21}x_1 + a_{22}x_2 + a_{23}x_3 \ )\mathbf{j} \\
&\quad + (\ a_{31}x_1 + a_{32}x_2 + a_{33}x_3 \ )\mathbf{k} \\
&= (\ y_1, y_2, y_3 \ ).
\end{aligned}
$$

In matrix notation this transformation will be written

$$
\begin{pmatrix} a_{11} & a_{12} & a_{13} \\ a_{21} & a_{22} & a_{23} \\ a_{31} & a_{32} & a_{33} \end{pmatrix} \begin{pmatrix} x_1 \\ x_2 \\ x_3 \end{pmatrix} = (\ y_1, y_2, y_3 \ ),
$$

where the multiplication indicated is the sum of the products of elements across each row and down the column vector.

DEFINITION 12. The *dot product* of two vectors in R^n is a real number that is the sum of the products of the corresponding components of the given vectors

$$\mathbf{X} \cdot \mathbf{Y} = x_1 y_1 + x_2 y_2 + \ldots + x_n y_n.$$

Let **A** stand for the matrix of coefficients above, called the *coefficient matrix* of the given linear transformation. Then in the transformed vector $y_1 = a_{11}x_1 + a_{12}x_2 + a_{13}x_3$ is the dot product of the first row of **A** with **X**, y_2 is the dot product of the second row of **A** with **X**, and y_3 is the dot product of the third row of **A** with **X**. In general the columns of matrix **A** are made up of the m-tuples found by transforming the basic vectors in R^n with the standard basis. Thus the first column of **A** is $T(i)$, the second column of **A** is $T(j)$, and the third column is $T(k)$. The pattern is continued for n columns in space R^n. The number of rows in matrix **A** is determined by the number of components in vector **Y**.

It is also possible to start with a given matrix **A** of real numbers and define a corresponding linear transformation. Take the dot product of each row with a given vector **X** in R^n to be the corresponding component of the transformed vector **Y**. The number of columns of **A** determines the number of components in **X**. If there are n columns the set of all n-tuple vectors to be transformed is called the *domain* vector space R^n. The number of rows of **A**, m, determines the number of components in **Y**. The collection of all these transformed vectors is called the *range* vector space R^m of the linear transformation. It is not difficult to show that the mapping described here determines a one-to-one correspondence between the set of all linear transformations of R^n to R^m and the set of all real m by n matrices. This means that for each linear transformation there is a unique matrix and for each matrix there is a corresponding unique linear transformation.

Some special cases are given in the following examples. In the 1 by 1 case, **A** = a for any real number a and

$$T(\mathbf{X}) = T(x) = ax = y = \mathbf{Y}$$

for every vector **X** in R^1. Test $F(x) = ax+b$ and show that F is not linear if $b \neq 0$.

In the case R^n to R^1,

$$\mathbf{A} = (a_{11}\ a_{12} \cdots a_{1n})$$

and

$$T(\mathbf{X}) = (a_{11}\ a_{12} \cdots a_{1n}) \begin{pmatrix} x_1 \\ x_2 \\ \cdot \\ \cdot \\ \cdot \\ x_n \end{pmatrix}$$

$$= a_{11}x_1 + a_{12}x_2 + \cdots + a_{1n}x_n = y = \mathbf{Y}.$$

Y is simply the dot product of vectors **A** and **X**.

In the case R^1 to R^m,

$$
\mathbf{A} = \begin{pmatrix} a_{11} \\ a_{21} \\ \cdot \\ \cdot \\ \cdot \\ a_{m1} \end{pmatrix}
$$

and

$$
T(\mathbf{X}) = \begin{pmatrix} a_{11} \\ a_{21} \\ \cdot \\ \cdot \\ \cdot \\ a_{m1} \end{pmatrix} \quad x = \begin{pmatrix} a_{11}x \\ a_{21}x \\ \cdot \\ \cdot \\ \cdot \\ a_{m1}x \end{pmatrix} = (y_1, y_2, \ldots, y_m) = \mathbf{Y}.
$$

For an example of R^3 to R^3 suppose $T(\mathbf{X})$ rotates every vector in R^3 90° counterclockwise about the z axis so that the x axis is carried into the y axis. Then

$$
\begin{aligned}
T(\mathbf{i}) &= \mathbf{j} &&= (0, 1, 0) \\
T(\mathbf{j}) &= -\mathbf{i} &&= (-1, 0, 0) \\
T(\mathbf{k}) &= \mathbf{k} &&= (0, 0, 1).
\end{aligned}
$$

$$
T(\mathbf{X}) = \mathbf{AX} = \begin{pmatrix} 0 & -1 & 0 \\ 1 & 0 & 0 \\ 0 & 0 & 1 \end{pmatrix} \begin{pmatrix} x_1 \\ x_2 \\ x_3 \end{pmatrix} = (-x_2, x_1, x_3) = \mathbf{Y}
$$

In particular $T(1, 2, 3) = (-2, 1, 3)$ which is the 90° rotation of $(1, 2, 3)$. Rotations of axes are linear transformations while translations of axes are not.

The sum of two linear transformations may be defined so that the sum is linear.

Definition 13.   If T and F are both linear transformations from R^n to R^m, then $T + F$ is the sum of the transformed vectors in R^m.

$$(T + F)(\mathbf{X}) = T(\mathbf{X}) + F(\mathbf{X})$$

Theorem 1-1. $T + F$ is linear if both T and F are linear transformations from R^n to R^m.

PROOF:

$$(T + F)(\alpha \mathbf{X}) = T(\alpha \mathbf{X}) + F(\alpha \mathbf{X})$$

$$= \alpha T(\mathbf{X}) + \alpha F(\mathbf{X})$$

$$= \alpha [\, T(\mathbf{X}) + F(\mathbf{X}) \,]$$

$$= \alpha (\, T + F \,)(\mathbf{X})$$

$$(T + F)(\mathbf{X} + \mathbf{Y}) = T(\mathbf{X} + \mathbf{Y}) + F(\mathbf{X} + \mathbf{Y})$$

$$= T(\mathbf{X}) + T(\mathbf{Y}) + F(\mathbf{X}) + F(\mathbf{Y})$$

$$= T(\mathbf{X}) + F(\mathbf{X}) + T(\mathbf{Y}) + F(\mathbf{Y})$$

$$= (\, T + F \,)(\mathbf{X}) + (\, T + F \,)(\mathbf{Y}). \quad \blacksquare [1]$$

If $\mathbf{A}$ is the matrix of T and $\mathbf{B}$ is the matrix of F then $\mathbf{A} + \mathbf{B}$ is defined to be the matrix of $T + F$. The jth column vector of $\mathbf{A} + \mathbf{B}$ is the transform of the jth basic $\mathbf{i}_j$ in R^n.

$$(\, T + F \,)(\, \mathbf{i}_j \,) = T(\, \mathbf{i}_j \,) + F(\, \mathbf{i}_j \,)$$

which is the vector sum of the jth column of $\mathbf{A}$ and the jth column of $\mathbf{B}$. Thus $\mathbf{A} + \mathbf{B}$ is the matrix whose columns are the vector sums of the corresponding columns of $\mathbf{A}$ and $\mathbf{B}$. In other words every element of $\mathbf{A} + \mathbf{B}$ is the sum of the two corresponding elements of $\mathbf{A}$ and $\mathbf{B}$. For example

$$\begin{pmatrix} 1 & 2 & 3 \\ 4 & 5 & 6 \end{pmatrix} + \begin{pmatrix} 2 & -1 & 0 \\ -3 & 0 & -4 \end{pmatrix} = \begin{pmatrix} 3 & 1 & 3 \\ 1 & 5 & 2 \end{pmatrix}$$

For the addition of two matrices to be defined, it is necessary that both matrices have the same size, that is, both are m by n.

We now know how to add matrices. Let's see how to multiply a matrix by a scalar.

DEFINITION 14.   The product αT for real α is the transformation that sends $\mathbf{X}$ into $\alpha T(\mathbf{X})$,

$$(\alpha T)(\mathbf{X}) = \alpha T(\mathbf{X}).$$

THEOREM 1-2. αT *is linear if T is a linear transformation.*

PROOF: $$(\alpha T)(\beta \mathbf{X}) = \alpha T(\beta \mathbf{X})$$

$$= \alpha \beta T(\mathbf{X})$$

$$= \beta \alpha T(\mathbf{X})$$

---

[1] $\blacksquare$ indicates the end of a proof.

$$= \beta(\,\alpha T\,)(\,\mathbf{X}\,)$$
$$(\,\alpha T\,)(\,\mathbf{X} + \mathbf{Y}\,) = \alpha T(\,\mathbf{X} + \mathbf{Y}\,)$$
$$= \alpha[\,T(\,\mathbf{X}\,) + T(\,\mathbf{Y}\,)\,]$$
$$= \alpha T(\,\mathbf{X}\,) + \alpha T(\,\mathbf{Y}\,)$$
$$= (\,\alpha T\,)(\,\mathbf{X}\,) + (\,\alpha T\,)(\,\mathbf{Y}\,). \quad \blacksquare$$

If $\mathbf{A}$ is the matrix of T then, as expected, $\alpha\mathbf{A}$ is defined to be the matrix of αT. Since

$$(\,\alpha T\,)\mathbf{X} = \alpha T(\,\mathbf{X}\,) = \alpha\mathbf{Y} = (\,\alpha y_1, \alpha y_2, \ldots, \alpha y_m\,)$$

$\alpha\mathbf{A}$ is found by multiplying each of the rows of $\mathbf{A}$ by α. For example, if $\alpha = 4$ and $T(\,\mathbf{i}\,) = (\,1, 4\,)$, $T(\,\mathbf{j}\,) = (\,2, 5\,)$, $T(\,\mathbf{k}\,) = (\,3, 6\,)$, then

$$(\,\alpha T\,)(\,\mathbf{X}\,) = 4 \begin{pmatrix} 1 & 2 & 3 \\ 4 & 5 & 6 \end{pmatrix} \begin{pmatrix} x_1 \\ x_2 \\ x_3 \end{pmatrix}$$

$$= 4 \begin{pmatrix} x_1 + 2x_2 + 3x_3 \\ 4x_1 + 5x_2 + 6x_3 \end{pmatrix}$$

$$= \begin{pmatrix} 4x_1 + 8x_2 + 12x_3 \\ 16x_1 + 20x_2 + 24x_3 \end{pmatrix}$$

$$= \begin{pmatrix} 4 & 8 & 12 \\ 16 & 20 & 24 \end{pmatrix} \begin{pmatrix} x_1 \\ x_2 \\ x_3 \end{pmatrix} = (\,\alpha\mathbf{A}\,)(\,\mathbf{X}\,).$$

Thus

$$4 \begin{pmatrix} 1 & 2 & 3 \\ 4 & 5 & 6 \end{pmatrix} = \begin{pmatrix} 4 & 8 & 12 \\ 16 & 20 & 24 \end{pmatrix}$$

and every element of $\mathbf{A}$ is multiplied by α.

To see how matrices multiply consider the composite of two linear transformations. If $T(\,\mathbf{X}\,)$ is a linear mapping of R^p to R^m,

$$T(\,\mathbf{X}\,) = T(\,x_1, x_2, \cdots, x_p\,) = (\,y_1, y_2, \cdots, y_m\,) = \mathbf{Y},$$

and $F(\mathbf{Y})$ is a linear mapping of R^m to R^q,

$$F(y_1, y_2, \cdots, y_m) = (z_1, z_2, \cdots, z_q) = \mathbf{Z},$$

then $F(T(\mathbf{X}))$ maps R^p to R^q by first applying T followed by F, called the *composite* mapping.

THEOREM 1-3. *$F(T(\mathbf{X}))$ is linear if both transformations F and T are linear.*

PROOF:
$$F(T(\alpha\mathbf{X})) = F(\alpha T(\mathbf{X}))$$
$$= \alpha F(T(\mathbf{X}))$$
$$F(T(\mathbf{X+Y})) = F(T(\mathbf{X}) + T(\mathbf{Y}))$$
$$= F(T(\mathbf{X})) + F(T(\mathbf{Y})). \qquad \blacksquare$$

By finding the matrix of the composite mapping, we will find the product of the two matrices corresponding to F and T. Let $\mathbf{B}$ be the m by p matrix corresponding to T. Then

$$T(\mathbf{X}) = \mathbf{BX}$$

$$= \begin{pmatrix} b_{11} & b_{12} & \cdots & b_{1p} \\ b_{21} & b_{22} & \cdots & b_{2p} \\ \cdot & \cdot & & \cdot \\ \cdot & \cdot & & \cdot \\ \cdot & \cdot & & \cdot \\ b_{m1} & b_{m2} & \cdots & b_{mp} \end{pmatrix} \begin{pmatrix} x_1 \\ x_2 \\ \cdot \\ \cdot \\ \cdot \\ x_p \end{pmatrix}$$

$$= \begin{pmatrix} b_{11}x_1 + b_{12}x_2 + \cdots + b_{1p}x_p \\ b_{21}x_1 + b_{22}x_2 + \cdots + b_{2p}x_p \\ \cdot \qquad \cdot \qquad \cdot \\ \cdot \qquad \cdot \qquad \cdot \\ \cdot \qquad \cdot \qquad \cdot \\ b_{m1}x_1 + b_{m2}x_2 + \cdots + b_{mp}x_p \end{pmatrix}$$

$$= (y_1, y_2, \cdots, y_m) = \mathbf{Y}.$$

Let $\mathbf{A}$ be the q by m matrix corresponding to F. Then

$F(\mathbf{Y}) = \mathbf{AY}$

$$= \begin{pmatrix} a_{11}a_{12} \cdots a_{1m} \\ a_{21}a_{22} \cdots a_{2m} \\ \cdot \quad \cdot \qquad \cdot \\ \cdot \quad \cdot \qquad \cdot \\ \cdot \quad \cdot \qquad \cdot \\ a_{q1}a_{q2} \ldots a_{qm} \end{pmatrix} \begin{pmatrix} y_1 \\ y_2 \\ \cdot \\ \cdot \\ \cdot \\ y_m \end{pmatrix}$$

$$= \begin{pmatrix} a_{11}y_1 + a_{12}y_2 + \cdots + a_{1m}y_m \\ a_{21}y_1 + a_{22}y_2 + \cdots + a_{2m}y_m \\ \cdot \qquad \cdot \qquad \qquad \cdot \\ \cdot \qquad \cdot \qquad \qquad \cdot \\ \cdot \qquad \cdot \qquad \qquad \cdot \\ a_{q1}y_1 + a_{q2}y_2 + \cdots + a_{qm}y_m \end{pmatrix}$$

$$= (z_1, z_2, \cdots, z_q) = \mathbf{Z}.$$

Composing,

$F(T(\mathbf{X})) = \mathbf{A}(\mathbf{BX})$

$$= \begin{pmatrix} a_{11}a_{12} \cdots a_{1m} \\ a_{21}a_{22} \cdots a_{2m} \\ \cdot \quad \cdot \qquad \cdot \\ \cdot \quad \cdot \qquad \cdot \\ \cdot \quad \cdot \qquad \cdot \\ a_{q1}a_{q2} \cdots a_{qm} \end{pmatrix} \begin{pmatrix} b_{11}b_{12} \cdots b_{1p} \\ b_{21}b_{22} \cdots b_{2p} \\ \cdot \quad \cdot \qquad \cdot \\ \cdot \quad \cdot \qquad \cdot \\ \cdot \quad \cdot \qquad \cdot \\ b_{m1}b_{m2} \cdots b_{mp} \end{pmatrix} \begin{pmatrix} x_1 \\ x_2 \\ \cdot \\ \cdot \\ \cdot \\ x_p \end{pmatrix}$$

$$= \begin{pmatrix} a_{11}a_{12} \cdots a_{1m} \\ a_{21}a_{22} \cdots a_{2m} \\ \cdot \quad \cdot \qquad \cdot \\ \cdot \quad \cdot \qquad \cdot \\ \cdot \quad \cdot \qquad \cdot \\ a_{q1}a_{q2} \cdots a_{qm} \end{pmatrix} \begin{pmatrix} b_{11}x_1 + b_{12}x_2 + \cdots + b_{1p}x_p \\ b_{21}x_1 + b_{22}x_2 + \cdots + b_{2p}x_p \\ \cdot \qquad \cdot \qquad \qquad \cdot \\ \cdot \qquad \cdot \qquad \qquad \cdot \\ \cdot \qquad \cdot \qquad \qquad \cdot \\ b_{m1}x_1 + b_{m2}x_2 + \cdots + b_{mp}x_p \end{pmatrix}$$

$$= \begin{pmatrix} a_{11} \sum_1^p b_{1j} x_j + a_{12} \sum_1^p b_{2j} x_j + \cdots + a_{1m} \sum_1^p b_{mj} x_j \\ a_{21} \sum_1^p b_{1j} x_j + a_{22} \sum_1^p b_{2j} x_j + \cdots + a_{2m} \sum_1^p b_{mj} x_j \\ \vdots \qquad\qquad \vdots \qquad\qquad\qquad \vdots \\ a_{q1} \sum_1^p b_{1j} x_j + a_{q2} \sum_1^p b_{2j} x_j + \cdots + a_{qm} \sum_1^p b_{mj} x_j \end{pmatrix}^1$$

$$= \begin{pmatrix} \sum_{i=1}^m a_{1i} \sum_{j=1}^p b_{ij} x_j \\ \sum_{i=1}^m a_{2i} \sum_{j=1}^p b_{ij} x_j \\ \vdots \\ \sum_{i=1}^m a_{qi} \sum_{j=1}^p b_{ij} x_j \end{pmatrix}$$

These finite sums may be regrouped to interchange the order of the summation and the common x_j factored out of each group of p terms. The result is the form given next for this matrix.

$$= \begin{pmatrix} \sum_{j=1}^p x_j \sum_{i=1}^m a_{1i} b_{ij} \\ \sum_{j=1}^p x_j \sum_{i=1}^m a_{2i} b_{ij} \\ \vdots \\ \sum_{j=1}^p x_j \sum_{i=1}^m a_{qi} b_{ij} \end{pmatrix}$$

By bringing out x_j as a column vector, an array of q rows and p columns is left, as shown in the next matrix.

$$= \begin{pmatrix} \sum_1^m a_{1i} b_{i1} & \sum_1^m a_{1i} b_{i2} & \cdots & \sum_1^m a_{1i} b_{ip} \\ \sum_1^m a_{2i} b_{i1} & \sum_1^m a_{2i} b_{i2} & \cdots & \sum_1^m a_{2i} b_{ip} \\ \vdots & \vdots & & \vdots \\ \sum_1^m a_{qi} b_{i1} & \sum_1^m a_{qi} b_{i2} & \cdots & \sum_1^m a_{qi} b_{ip} \end{pmatrix} \begin{pmatrix} x_1 \\ x_2 \\ \vdots \\ x_p \end{pmatrix}$$

This final array, q by p, is the matrix **AB** that transforms **X** into **Z**. By comparing the product **AB** with the original matrices **A** and **B**, we can see how to define the multiplication. The element in the jth row and k^{th} column is

$$\sum_{i=1}^m a_{ji} b_{ik}.$$

This formula adds the products of the corresponding elements across the jth row of **A** and down the kth column of **B**. The jk^{th} element in the matrix **AB** is thus

---

[1] $\sum_1^p b_{1j} x_j = b_{11} x_1 + b_{12} x_2 + \cdots + b_{1p} x_p.$

defined to be the dot product of the vector in the jth row of **A** with the vector in the kth column of **B**. These two vectors must have the same number of coordinates for their dot product to be defined. They have the same number of coordinates because the number of columns in **A** equals the number of rows in **B**.

Example 1:

$$\begin{pmatrix} 1 & 2 & 3 \\ 4 & 5 & 6 \\ 7 & 8 & 9 \\ 0 & 1 & 0 \end{pmatrix} \begin{pmatrix} 2 & 1 \\ -1 & 0 \\ 0 & -2 \end{pmatrix} = \begin{pmatrix} 1(2) + 2(-1) + 3(0) & 1(1) + 2(0) + 3(-2) \\ 4(2) + 5(-1) + 6(0) & 4(1) + 5(0) + 6(-2) \\ 7(2) + 8(-1) + 9(0) & 7(1) + 8(0) + 9(-2) \\ 0(2) + 1(-1) + 0(0) & 0(1) + 1(0) + 0(-2) \end{pmatrix}$$

$$= \begin{pmatrix} 0 & -5 \\ 3 & -8 \\ 6 & -11 \\ -1 & 0 \end{pmatrix}$$

●[2]

Since the number of columns of **A** must agree with the number of rows of **B**, the product **AB** is not, in general, commutative. However, matrix multiplication may be shown to be associative and distributive over sums wherever all the products involved are defined.

A(**BC**) = (**AB**)C

A(**B** + **C**) = **AB** + **AC**

(**A** + **B**)C = **AC** + **BC**

These follow from the similar laws of the corresponding linear transformations. As a matter of fact it is clear from our development that the algebra of matrices is equivalent to the algebra of their linear transformations.

1.4 IDENTITY AND INVERSE MAPS

DEFINITION 15. The *identity mapping* in R^n is a transformation that sends every point of R^n into itself, $T(\mathbf{X}) = \mathbf{X}$.

Let *I* stand for the identity mapping that is easily shown to satisfy both conditions of linearity. Since linear transformation *I* sends each of the basic vectors into itself, the matrix of *I* is made up of columns that are the respective unit vectors in the standard basis. For R^3

---

[2] ● indicates the end of an example.

$$I(\mathbf{X}) = \begin{pmatrix} 1 & 0 & 0 \\ 0 & 1 & 0 \\ 0 & 0 & 1 \end{pmatrix} \begin{pmatrix} x_1 \\ x_2 \\ x_3 \end{pmatrix} = (x_1, x_2, x_3).$$

The matrix of mapping I, called the *identity matrix*, has ones down the main diagonal and zeros elsewhere. The identity matrix, $\mathbf{I}^n$, is square and n by n since R^n has n basic vectors. For the standard basis these basic vectors have one in the ith place and zeros in the remaining $n - 1$ places for $i = 1, 2, \ldots, n$.

DEFINITION 16. A linear transformation T has an *inverse* called T^{-1} if the composite mappings TT^{-1} and $T^{-1}T$ are both the identity mapping. When this occurs we say that T is *nonsingular*.

If $T(\mathbf{X}) = \mathbf{Y}$ maps R^n to R^m then $T^{-1}(\mathbf{Y}) = \mathbf{X}$ maps R^m to R^n. Composing these maps we have $TT^{-1}(\mathbf{Y}) = T(\mathbf{X}) = \mathbf{Y}$ is an identity mapping of R^m to R^m and $T^{-1}T(\mathbf{X}) = T^{-1}(\mathbf{Y}) = (\mathbf{X})$ is an identity mapping of R^n to R^n. It would appear that there are two identity mappings involved but the inverse is unique and exists only if $n = m$. Then

$$TT^{-1} = T^{-1}T = I.$$

THEOREM 1-4. *When it exists the inverse mapping is unique.*

PROOF: Suppose F and G are both inverses of T. Then
$$FT = TF = I \quad \text{and}$$
$$GT = TG = I.$$
$$F = IF = (GT)F = G(TF) = GI = G. \quad \blacksquare$$

Thus F and G are the same so there is only one inverse.

THEOREM 1-5. *The inverse mapping, if it exists, is linear.*

PROOF: Suppose $T^{-1}T = TT^{-1} = I$.
Let $T^{-1}(\alpha\mathbf{X}) = \mathbf{Y}$ and $\alpha T^{-1}(\mathbf{X}) = \mathbf{Z}$.
Then $TT^{-1}(\alpha\mathbf{X}) = T(\mathbf{Y})$, $\quad\quad T(\alpha T^{-1}(\mathbf{X})) = T(\mathbf{Z})$
$\quad\quad\quad I(\alpha\mathbf{X}) = T(\mathbf{Y}) \quad\quad\quad\quad \alpha TT^{-1}(\mathbf{X}) = T(\mathbf{Z})$
$\quad\quad\quad\quad \alpha\mathbf{X} = T(\mathbf{Y}) \quad\quad\quad\quad\quad \alpha I(\mathbf{X}) = T(\mathbf{Z})$
$\quad\quad\quad\quad\quad\quad\quad\quad\quad\quad\quad\quad\quad\quad \alpha\mathbf{X} = T(\mathbf{Z})$, so $T(\mathbf{Y}) = T(\mathbf{Z})$.
Since T^{-1} has a unique image by theorem 1-4, $\mathbf{Y} = \mathbf{Z}$, completing the first condition.

Let $T^{-1}(\mathbf{X} + \mathbf{Y}) = \mathbf{Z}$ and $T^{-1}(\mathbf{X}) + T^{-1}(\mathbf{Y}) = \mathbf{W}$.
Then: $TT^{-1}(\mathbf{X} + \mathbf{Y}) = T(\mathbf{Z})$, $\quad\quad T(T^{-1}(\mathbf{X}) = T^{-1}(\mathbf{Y})) = T(\mathbf{W})$
$\quad\quad\quad I(\mathbf{X} + \mathbf{Y}) = T(\mathbf{Z}) \quad\quad\quad TT^{-1}(\mathbf{X}) + TT^{-1}(\mathbf{Y}) = T(\mathbf{W})$
$\quad\quad\quad\quad\quad \mathbf{X} + \mathbf{Y} = T(\mathbf{Z}) \quad\quad\quad\quad\quad\quad I(\mathbf{X}) + I(\mathbf{Y}) = T(\mathbf{W})$
$\quad\quad\quad\quad\quad\quad\quad\quad\quad\quad\quad\quad\quad\quad\quad\quad\quad \mathbf{X} + \mathbf{Y} = T(\mathbf{W})$,
$\quad\quad\quad\quad\quad\quad\quad\quad\quad\quad\quad\quad\quad\quad\quad\quad$ so $T(\mathbf{Z}) = T(\mathbf{W})$.
Again by the uniqueness of T^{-1}, $\mathbf{Z} = \mathbf{W}$, completing the proof of linearity. $\quad \blacksquare$

Define $\mathbf{A}^{-1}$, called $\mathbf{A}$ *inverse*, to be the matrix of linear transformation T^{-1}. We then ask how to find $\mathbf{A}^{-1}$ from $\mathbf{A}$ in order to preserve the product rule $\mathbf{A}^{-1}\mathbf{A} = \mathbf{A}\mathbf{A}^{-1} = \mathbf{I}^n$ where $\mathbf{I}^n$ is the n by n identity matrix. Let $T(\mathbf{X}) = \mathbf{A}\mathbf{X}$; map R^n to R^n and let the basic unit vectors of R^n be $\mathbf{i}_1, \mathbf{i}_2, \ldots, \mathbf{i}_n$. The columns of $\mathbf{A}$ are $T(\mathbf{i}_1) = \mathbf{I}_1$, $T(\mathbf{i}_2) = \mathbf{I}_2, \ldots, T(\mathbf{i}_n) = \mathbf{I}_n$ where $\mathbf{I}_j = \Sigma_{k=1}^n a_{kj}\mathbf{i}_k$ for each $j = 1, 2, \ldots, n$. The idea is to solve this system of n linear equations for the $\mathbf{i}_k$ as linear combinations of the $\mathbf{I}_j$. If there is a unique solution,

$$\mathbf{i}_k = \Sigma_{j=1}^n b_{jk}\mathbf{I}_j \text{ for } k = 1, 2, \ldots, n.$$

These equations say that the columns of the product $\mathbf{AB}$ are linear combinations of the columns of $\mathbf{A}$ and the sum above gives the corresponding basic unit vector for each column, i.e., $\mathbf{AB} = \mathbf{I}^n$ and $\mathbf{B} = \mathbf{A}^{-1}$.

Example 2: Find the inverse of the rotation matrix that was derived earlier.

$$\mathbf{A} = \begin{pmatrix} 0 & -1 & 0 \\ 1 & 0 & 0 \\ 0 & 0 & 1 \end{pmatrix}$$

The transforms of the unit vectors are

$$\mathbf{I}_1 = 0\mathbf{i}_1 + 1\mathbf{i}_2 + 0\mathbf{i}_3$$
$$\mathbf{I}_2 = -1\mathbf{i}_1 + 0\mathbf{i}_2 + 0\mathbf{i}_3$$
$$\mathbf{I}_3 = 0\mathbf{i}_1 + 0\mathbf{i}_2 + 1\mathbf{i}_3, \text{ which are the columns of } \mathbf{A}.$$

Solving

$$\mathbf{i}_1 = 0\mathbf{I}_1 - 1\mathbf{I}_2 + 0\mathbf{I}_3$$
$$\mathbf{i}_2 = 1\mathbf{I}_1 + 0\mathbf{I}_2 + 0\mathbf{I}_3$$
$$\mathbf{i}_3 = 0\mathbf{I}_1 + 0\mathbf{I}_2 + 1\mathbf{I}_3$$

The coefficients of the transforms, $\mathbf{I}_j$, are the columns of $\mathbf{A}^{-1}$, where

$$\mathbf{i}_1 = T^{-1}(\mathbf{I}_1), \mathbf{i}_2 = T^{-1}(\mathbf{I}_2), \mathbf{i}_3 = T^{-1}(\mathbf{I}_3).$$

As a check, multiply $\mathbf{A}$ by its inverse and see if the product is the identity matrix.

$$\mathbf{A}^{-1}\mathbf{A} = \begin{pmatrix} 0 & 1 & 0 \\ -1 & 0 & 0 \\ 0 & 0 & 1 \end{pmatrix}\begin{pmatrix} 0 & -1 & 0 \\ 1 & 0 & 0 \\ 0 & 0 & 1 \end{pmatrix} = \begin{pmatrix} 1 & 0 & 0 \\ 0 & 1 & 0 \\ 0 & 0 & 1 \end{pmatrix} = \mathbf{I}^3$$

●

In the next section a simpler method is found for arriving at the inverse matrix.

THEOREM 1-6. *If* $\mathbf{A}^{-1}$ *exists then* $(\ \mathbf{A}^{-1}\ )^{-1} = \mathbf{A}$.

PROOF: Since $\mathbf{A}^{-1}\mathbf{A} = \mathbf{A}\mathbf{A}^{-1} = \mathbf{I}$, $\mathbf{A}$ is the inverse of $\mathbf{A}^{-1}$.   ■

THEOREM 1-7. *If matrices* $\mathbf{A}$ *and* $\mathbf{B}$ *have inverses and the product* $\mathbf{AB}$ *is defined then*

$$(\ \mathbf{AB}\ )^{-1} = \mathbf{B}^{-1}\mathbf{A}^{-1}$$

PROOF:

$$(\ \mathbf{B}^{-1}\mathbf{A}^{-1}\ )(\ \mathbf{AB}\ ) = \mathbf{B}^{-1}(\ \mathbf{A}^{-1}\mathbf{A}\ )\mathbf{B}\ ) = \mathbf{B}^{-1}\mathbf{I}^{n}\mathbf{B} = \mathbf{B}^{-1}\mathbf{B} = \mathbf{I}^{n}.$$

and $\mathbf{AB}(\ \mathbf{B}^{-1}\mathbf{A}^{-1}\ ) = \mathbf{A}(\ \mathbf{BB}^{-1}\ )\mathbf{A}^{-1} = \mathbf{AI}^{n}\mathbf{A}^{-1} = \mathbf{AA}^{-1} = \mathbf{I}^{n}.$ ■

1.5 SOLVING LINEAR EQUATIONS

Consider a set of m linear equations in n unknowns

$$a_{11}x_1 + a_{12}x_2 + \cdots + a_{1n}x_n = y_1$$
$$a_{21}x_1 + a_{22}x_2 + \cdots + a_{2n}x_n = y_2$$

$$\vdots \qquad\qquad\qquad\qquad\qquad (1)$$

$$a_{m1}x_1 + a_{m2}x_2 + \cdots + a_{mn}x_n = y_m$$

System (1) is equivalent to the linear transformation $T(\ \mathbf{X}\ ) = \mathbf{Y}$ from R^n to R^m where vector $\mathbf{X}$ is unknown and vector $\mathbf{Y}$ along with matrix $\mathbf{A}$ is given. The matrix of the linear transformation is the array of coefficients in (1).

$$\mathbf{A} = \begin{pmatrix} a_{11}a_{12} \cdots a_{1n} \\ a_{21}a_{22} \cdots a_{2n} \\ \cdot \quad \cdot \qquad \cdot \\ \cdot \quad \cdot \qquad \cdot \\ \cdot \quad \cdot \qquad \cdot \\ a_{m1}a_{m2} \cdots a_{mn} \end{pmatrix}$$

DEFINITION 17. System (1) written as a matrix equation is $\mathbf{AX} = \mathbf{Y}$ where $\mathbf{A}$ will be called the *coefficient matrix* and vector $\mathbf{Y}$ the set of *constant terms*. A *solution* to (1) is any vector $\mathbf{X}$ that satisfies all m equations.

DEFINITION 18. The *augmented matrix* is the m by $n + 1$ matrix made up of $\mathbf{A}$ plus the column of constant terms.

$$\begin{pmatrix} a_{11}a_{12} \cdots a_{1n}y_1 \\ a_{21}a_{22} \cdots a_{2n}y_2 \\ \cdot \quad \cdot \quad\quad \cdot \quad \cdot \\ \cdot \quad \cdot \quad\quad \cdot \quad \cdot \\ \cdot \quad \cdot \quad\quad \cdot \quad \cdot \\ a_{m1}a_{m2} \cdots a_{mn}y_m \end{pmatrix}$$

Solution vectors **X** for system (1) will be found by an elimination process known as *Gauss-Jordan* elimination.

Gauss-Jordan elimination uses the principle that any multiple of one equation may be added to or subtracted from another. The idea relies upon the axioms that equals multiplied by equals are equal and equals added to equals are equal. Subtraction is accomplished by multiplying by a negative number and then adding. The order of the equations in (1) is immaterial, so three operations are sufficient in the elimination process; they are called *elementary row operations*:

1. Interchanging any two rows.
2. Multiplying any row by a nonzero constant.
3. Adding a multiple of one row to any other row.

During the elimination process nothing is gained by carrying along the unknowns or the equal signs. The writing is shortened by applying these three operations directly to the augmented matrix to reduce it to the augmented matrix of an equivalent system that is easy to solve.

DEFINITION 19. An *equivalent system* to (1) is a system of linear equations that has exactly the same set of solution vectors as (1).

The three elementary row operations just discussed, when applied to any system of linear equations, necessarily reduce it to an equivalent system.

Example 3: $2x_1 - x_2 + 2x_3 = 6$

$$x_1 + 2x_2 - x_3 = 2$$

$$3x_1 - x_2 - 3x_3 = -8$$

The 3 by 4 augmented matrix is

$$\begin{pmatrix} 2 & -1 & 2 & 6 \\ 1 & 2 & -1 & 2 \\ 3 & -1 & -3 & -8 \end{pmatrix}$$

The Gauss-Jordan elimination begins by finding, if possible, a nonzero entry in the first column. Using operation 1 this entry may be put into the first row. If there are no nonzero entries in column one, we move to the second column and proceed as if it were the first column. Next use operation 2 to produce a 1 in the a_{11} position, i.e., multiply row one by the reciprocal of its entry in the first column.

$$\begin{pmatrix} 1 & -\frac{1}{2} & 1 & 3 \\ 1 & 2 & -1 & 2 \\ 3 & -1 & -3 & -8 \end{pmatrix}$$

By operation 3 all remaining entries in column one may be made zero. In this example subtract the first row from the second, and subtract 3 times the first row from the third.

$$\begin{pmatrix} 1 & -\frac{1}{2} & 1 & 3 \\ 0 & \frac{5}{2} & -2 & -1 \\ 0 & \frac{1}{2} & -6 & -17 \end{pmatrix}$$

Moving to column two we produce a 1 in the second row, provided there is a nonzero entry in the second column below the first row. Otherwise we move down the line until a column is found that satisfies this condition. In our case multiply the second row by $\frac{2}{5}$.

$$\begin{pmatrix} 1 & -\frac{1}{2} & 1 & 3 \\ 0 & 1 & -\frac{4}{5} & -\frac{2}{5} \\ 0 & \frac{1}{2} & -6 & -17 \end{pmatrix}$$

Again produce zeros in this column except for the 1 just obtained. Let $\frac{1}{2}$ times the second row be added to the first row and be subtracted from the third row.

$$\begin{pmatrix} 1 & 0 & \frac{3}{5} & \frac{14}{5} \\ 0 & 1 & -\frac{4}{5} & -\frac{2}{5} \\ 0 & 0 & -\frac{28}{5} & -\frac{84}{5} \end{pmatrix}$$

Move over another column and try to produce a 1 in the next row. Multiplying row three by $-\frac{5}{28}$ does the job.

$$\begin{pmatrix} 1 & 0 & \frac{3}{5} & \frac{14}{5} \\ 0 & 1 & -\frac{4}{5} & -\frac{2}{5} \\ 0 & 0 & 1 & 3 \end{pmatrix}$$

The example is finished by subtracting ³⁄₅ of the third row from the first and adding ⁴⁄₅ of the third row to the second.

$$\begin{pmatrix} 1 & 0 & 0 & 1 \\ 0 & 1 & 0 & 2 \\ 0 & 0 & 1 & 3 \end{pmatrix}$$

This system is equivalent to the original system and has its solution read off immediately from each row.

$$x_1 = 1, \ x_2 = 2, \ x_3 = 3. \quad \bullet$$

The completely reduced augmented matrix has as many unit column vectors as possible with each of the 1s in a different row. Such a form is called *row echelon form*. Since the row echelon form of the augmented matrix was reached by using only the three allowed operations, the corresponding system of equations has the same solution as the original system of equations, that is, the two systems are equivalent.

The example just completed has a unique solution but there are other possibilities. If the last column in the row echelon form turned out to be one of the unit column vectors, then some nonzero row would be

$$0 \quad 0 \quad 0 \quad \ldots \quad 0 \quad 1.$$

In this case the system has no solution because a contradiction, $0 = 1$, has been reached. A system with no solutions is called *inconsistent*, whereas a *consistent* system has one or more solutions. An inconsistent system necessarily leads to a contradiction by having a distinct unit vector in the last column of the row echelon form. Infinitely many solutions occur in a consistent system when the number of unit column vectors in its row echelon form is less than the number of unknowns. The following example is the row echelon form of the augmented matrix of a system with four equations and five unknowns.

$$\begin{pmatrix} 1 & 2 & 0 & 3 & 0 & 4 \\ 0 & 0 & 1 & 2 & 0 & 5 \\ 0 & 0 & 0 & 0 & 1 & 6 \\ 0 & 0 & 0 & 0 & 0 & 0 \end{pmatrix}$$

Distinct unit vectors could not be produced in columns two, four, or six, because all of their entries are zero below the 1 of the unit vector in a previous column. If a row contains all zero entries, it is moved to the bottom of the matrix. This row echelon form is the simplest form for the augmented matrix by row operations and the number of unit vectors is unique for any matrix. In the given example the solutions are

$$x_1 = 4 - 2_2 - 3x_4$$

$$x_3 = 5 - 2x_4$$

$$x_5 = 6$$

where the unknowns corresponding to the unit column vectors are solved for in terms of the remaining unknowns. These remaining unknowns x_2 and x_4 may be assigned any arbitrary value in the solution vector.

If $x_2 = c_1$ and $x_4 = c_2$ for arbitrary constants c_1, c_2, then the solutions in vector notation are:

$$\begin{pmatrix} x_1 \\ x_2 \\ x_3 \\ x_4 \\ x_5 \end{pmatrix} = \begin{pmatrix} 4 \\ 0 \\ 5 \\ 0 \\ 6 \end{pmatrix} + c_1 \begin{pmatrix} -2 \\ 1 \\ 0 \\ 0 \\ 0 \end{pmatrix} + c_2 \begin{pmatrix} -3 \\ 0 \\ -2 \\ 1 \\ 0 \end{pmatrix}$$

1.6 RANK OF A MATRIX

The row echelon form for any matrix **A** has a unique number of distinct unit vectors. Thus this number can be used to define the rank of **A**.

DEFINITION 20.   The *rank* of any matrix **A** is the number of distinct unit column vectors in its row echelon form.

Every row in the row echelon form of **A** after the 1 in the last unit column vector must consist entirely of zeros, otherwise more unit column vectors could be found. Each of the 1s of these unit column vectors is in a different row so the number of nonzero rows is the same as the number of distinct unit column vectors and is therefore equal to the rank of **A**.

THEOREM 1-8.   *A system of linear equations has a solution if and only if the rank of its augmented matrix equals the rank of its coefficient matrix.*

PROOF: We have already noted that a system of linear equations is inconsistent if and only if the row echelon form of its augmented matrix has a distinct unit vector in its last column. This can happen if and only if the augmented matrix has one more unique column unit vector than the coefficient matrix, i.e., if and only if the rank of the augmented matrix is one more than the rank of the coefficient matrix. The only other possibility is that the ranks are equal and this is the case if and only if at least one solution exists.   ∎

THEOREM 1-9.   *A system of linear equations has a unique solution if and only if the ranks of the augmented matrix and the coefficient matrix and the number of unknowns are all equal.*

PROOF: By theorem 1-8 if the ranks are equal, at least one solution exists. If the number of unknowns is also equal to this rank, then there are as many distinct unit vectors as columns in the row echelon form of the coefficient matrix. This means that in row echelon form, each of the unknowns is solved for in terms of the constants in the last column of the reduced augmented matrix. Since row reduction leads to an equivalent system, all possible orderings of the unit vectors under row operations have the same solution.

The converse is also true: if the solution is unique, the two ranks are equal to the number of unknowns. If the unit vectors corresponding to n unknowns are placed in order, they form the identity matrix whose rank is n and equals the rank of the augmented matrix. ■

Several special results or corollaries follow from theorem 1-9.

COROLLARY 1-1. *If coefficient matrix **A** is n by n, then the corresponding linear system with n equations in n unknowns has a unique solution if and only if the rank of **A** is n.*

PROOF: If the rank of **A** is n, then the augmented matrix has rank at least n, but since the augmented matrix has only n rows, its rank cannot exceed n. Thus the ranks of both matrices are n and theorem 1-9 guarantees the conclusion. ■

COROLLARY 1-2. *If a linear system has a solution, then the number of arbitrary constants in the solution is equal to the number of unknowns minus the rank of **A**.*

PROOF: If the rank of **A** is equal to the number of unknowns, n, then the solution is unique by corollary 1-1 and there are zero constants. Otherwise, the rank of **A** is less than n, and the difference is the number of "free" unknowns in the solution, or the number of constants. ■

The number of arbitrary constants is referred to as the number of *degrees freedom* in the solution vector.

1.7 INVERSE OF A MATRIX BY PIVOTING

The row echelon form of an n by n matrix **A** of rank n will necessarily be the identity matrix $\mathbf{I}^n$, since there are exactly n distinct unit column vectors which are found in order. Also an n by n matrix **A** of rank n has an inverse because the corresponding system $\mathbf{AX} = \mathbf{Y}$ has a unique solution. Thus, to every vector **Y** there is a unique vector **X** or $\mathbf{X} = \mathbf{A}^{-1}\mathbf{Y}$, where $\mathbf{A}^{-1}$ is the matrix of the unique inverse transformation. We now wish to find an easier way to compute the inverse matrix.

DEFINITION 21. An *elementary matrix* is the result of carrying out operations 1, 2, or 3 on an identity matrix.

If rows one and three are interchanged in $\mathbf{I}^3$, then the elementary matrix formed is

$$\mathbf{E} = \begin{pmatrix} 0 & 0 & 1 \\ 0 & 1 & 0 \\ 1 & 0 & 0 \end{pmatrix}$$

Multiplying an arbitrary 3 by 3 matrix $\mathbf{A}$ by $\mathbf{E}$

$$\begin{pmatrix} 0 & 0 & 1 \\ 0 & 1 & 0 \\ 1 & 0 & 0 \end{pmatrix} \begin{pmatrix} a_{11} a_{12} a_{13} \\ a_{21} a_{22} a_{23} \\ a_{31} a_{32} a_{33} \end{pmatrix} = \begin{pmatrix} a_{31} a_{32} a_{33} \\ a_{21} a_{22} a_{23} \\ a_{11} a_{12} a_{13} \end{pmatrix}$$

performs the same interchange on $\mathbf{A}$. Operation 2 applied to $\mathbf{I}^n$ multiplies a row of $\mathbf{I}^n$ by a nonzero constant to get an elementary $\mathbf{E}$. Clearly multiplying an n by n matrix $\mathbf{A}$ by $\mathbf{E}$ does the same thing to the corresponding row of $\mathbf{A}$. To illustrate operation 3, take 3 times the first row of $\mathbf{I}^3$ and subtract the result from the second row. Then

$$\mathbf{EA} = \begin{pmatrix} 1 & 0 & 0 \\ -3 & 1 & 0 \\ 0 & 0 & 1 \end{pmatrix} \begin{pmatrix} a_{11} a_{12} a_{13} \\ a_{21} a_{22} a_{23} \\ a_{31} a_{32} a_{33} \end{pmatrix} = \begin{pmatrix} a_{11} & a_{12} & a_{13} \\ a_{21}-3a_{11} & a_{22}-3a_{12} & a_{23}-3a_{13} \\ a_{31} & a_{32} & a_{33} \end{pmatrix}$$

which is the same operation 3 applied to $\mathbf{A}$.

In general let the jth row of $\mathbf{B}$ be k times the ith row of m by n matrix $\mathbf{A}$ added to the jth row of $\mathbf{A}$ and let $\mathbf{E}$ be the corresponding elementary matrix found by doing the same to $\mathbf{I}^m$. In all other rows let $\mathbf{B} = \mathbf{A}$. To show $\mathbf{B} = \mathbf{EA}$ note first the product $\mathbf{EA}$ leaves all rows of $\mathbf{A}$ the same except the jth row, as that is the only row of $\mathbf{E}$ that differs from $\mathbf{I}^m$. So $\mathbf{B}$ and $\mathbf{EA}$ agree except possibly for the jth row. Let the rows of $\mathbf{I}^m$ be labeled $E_1, E_2, \cdots, E_m$ and the columns of $\mathbf{A}$ be labeled $A_1, A_2, \cdots, A_n$. Then using $(\cdot)$ for the dot product, the jth row of $\mathbf{B}$ may be written

$$(a_{j1}, a_{j2}, \cdots, a_{jn}) + k (a_{i1}, a_{i2}, \cdots, a_{in})$$
$$= (E_j \cdot A_1, E_j \cdot A_2, \cdots, E_j \cdot A_n) + k (E_i \cdot A_1, E_i \cdot A_2, \cdots, E_i \cdot A_n)$$
$$= [(E_j + kE_i) \cdot A_1, (E_j + kE_i) \cdot A_2, \cdots, (E_j + kE_i) \cdot A_n].$$

This expression is the dot product of the jth row of $\mathbf{E}$ with each column of $\mathbf{A}$ and is therefore the jth row of the product $\mathbf{EA}$. Thus the jth row of $\mathbf{B}$ and $\mathbf{EA}$ agree as well as all others. Similarly we may show $\mathbf{B} = \mathbf{EA}$ under operations 1 and 2 to justify the following theorem.

THEOREM 1-10.   *If matrix* **B** *is the result of operations 1, 2, or 3 on m by n matrix* **A**, *then* **B** = **EA** *where* **E** *is the elementary matrix formed by using the same operation on the identity* **I**^{m}.

THEOREM 1-11.   *Each elementary matrix has an inverse that is also an elementary matrix.*

PROOF: If **E** is the result of interchanging the ith and j^{th} rows of **I**$''$, then **EE** = **I**$''$ so **E** is its own inverse. If **E**$_1$ is $k \neq 0$ times the ith row of **I**$''$ and **E**$_2$ is $1/k$ times the ith row of **I**$''$, then using theorem 1–10, **E**$_2$**E**$_1$ = **E**$_1$**E**$_2$ = **I**$''$. If **E**$_3$ is the matrix found by adding k times the ith row of **I**$''$ to the jth row of **I**$''$ and **E**$_4$ is found by adding $-k$ times the ith row of **I**$''$ to the jth row of **I**$''$ where $i \neq j$, then **E**$_4$**E**$_3$ = **E**$_3$**E**$_4$ = **I**$''$. If $i = j$ the multiple $k+1$ is taken of the ith row and its inverse is covered in case two. Therefore every elementary matrix has an elementary inverse.   ∎

DEFINITION 22.   A matrix that has no inverse is called *singular* and a matrix with an inverse is called *nonsingular*.

A matrix **A** will be nonsingular if there is a matrix **A**$^{-1}$ such that

$$\mathbf{AA}^{-1} = \mathbf{A}^{-1}\mathbf{A} = \mathbf{I}^n.$$

Since this product in either order is the same identity matrix, **A** must be square and n by n. The existence of **A**$^{-1}$ also means the linear transformation **X** = **A**$^{-1}$**Y** and therefore **Y** = **AX** both have unique solutions. By corollary 1–1 to theorem 1–9, **A** will then be of rank n and its row echelon form will be **I**$''$. For such a matrix the inverse may be found as follows.

THEOREM 1-12.   *If a nonsingular matrix* **A** *is reduced to row echelon form by a sequence of the operations 1, 2, 3, then* **A**$^{-1}$ *may be found by applying the same sequence of operations to the identity* **I**$''$.

PROOF: Let **E**$_1$, **E**$_2$, $\cdots$, **E**$_k$ be the sequence of elementary matrices corresponding to the sequence of elementary row operations on nonsingular **A** that reduce it to its row echelon form **I**$''$. By use of theorem 1-10 on each product from right to left

$$\mathbf{I}^n = \mathbf{E}_k\mathbf{E}_{k-1} \cdots \mathbf{E}_2\mathbf{E}_1\mathbf{A}.$$

Then

$$\mathbf{I}^n\mathbf{A}^{-1} = \mathbf{E}_k\mathbf{E}_{k-1} \cdots \mathbf{E}_2\mathbf{E}_1\mathbf{A}\mathbf{A}^{-1}$$

or

$$\mathbf{A}^{-1} = \mathbf{E}_k\mathbf{E}_{k-1} \cdots \mathbf{E}_2\mathbf{E}_1\mathbf{I}^n$$

which is the statement of the theorem.   ∎

For example, to find the inverse of

$$\mathbf{A} = \begin{pmatrix} 1 & 2 & 3 \\ -1 & 0 & 1 \\ 2 & -1 & 0 \end{pmatrix}$$

we will reduce A to row echelon form and at the same time perform each of the operations on $\mathbf{I}^3$.

$$\begin{pmatrix} 1 & 2 & 3 & | & 1 & 0 & 0 \\ -1 & 0 & 1 & | & 0 & 1 & 0 \\ 2 & -1 & 0 & | & 0 & 0 & 1 \end{pmatrix} \longrightarrow \begin{pmatrix} 1 & 2 & 3 & | & 1 & 0 & 0 \\ 0 & 2 & 4 & | & 1 & 1 & 0 \\ 0 & -5 & -6 & | & -2 & 0 & 1 \end{pmatrix} \longrightarrow$$

$$\begin{pmatrix} 1 & 2 & 3 & | & 1 & 0 & 0 \\ 0 & 1 & 2 & | & 1/2 & 1/2 & 0 \\ 0 & 0 & 4 & | & 1/2 & 5/2 & 1 \end{pmatrix} \longrightarrow \begin{pmatrix} 1 & 0 & -1 & | & 0 & -1 & 0 \\ 0 & 1 & 2 & | & 1/2 & 1/2 & 0 \\ 0 & 0 & 1 & | & 1/8 & 5/8 & 1/4 \end{pmatrix} \longrightarrow$$

$$\begin{pmatrix} 1 & 0 & 0 & | & 1/8 & -3/8 & 1/4 \\ 0 & 1 & 0 & | & 1/4 & -3/4 & -1/2 \\ 0 & 0 & 1 & | & 1/8 & 5/8 & 1/4 \end{pmatrix}$$

The n by n matrix $\mathbf{A}$ is increased to n by $2n$ with the addition of $\mathbf{I}^n$. When producing a unit column vector, the elementary row operations are carried out across all $2n$ entries of each row. To get the unit vector in the first column, the entire first row of six elements is added to the second row, and twice the first row is subtracted from all six elements across the third row. Similarly the second and third unit vectors are found with each operation carried out across the entire row. The final matrix in this example shows $\mathbf{A}$ has been reduced to $\mathbf{I}^3$ and the same operations have produced $\mathbf{A}^{-1}$ to the right of the vertical line. Check the result by multiplication.

$$\mathbf{AA}^{-1} = \mathbf{I}^3$$

$$\begin{pmatrix} 1 & 2 & 3 \\ -1 & 0 & 1 \\ 2 & -1 & 0 \end{pmatrix} \begin{pmatrix} 1/8 & -3/8 & 1/4 \\ 1/4 & -3/4 & -1/2 \\ 1/8 & 5/8 & 1/4 \end{pmatrix} = \begin{pmatrix} 1 & 0 & 0 \\ 0 & 1 & 0 \\ 0 & 0 & 1 \end{pmatrix}$$

Producing a unit column vector is an important operation in linear programming and will be called *pivoting*. The number used to get the one in a unit column vector is called the pivot in the current matrix. In the last example we pivoted on pivot 1 in the first column to get the unit vector in that column. In the next column we pivoted on the 2 of the second row to get the second unit vector. Finally, pivoting on 4 in the third row, third column gave the last unit vector. Finding $\mathbf{A}^{-1}$ is equivalent to pivoting down the main diagonal of A and at the same time carrying out these steps on $\mathbf{I}^n$. If a zero is encountered at a pivot position, then operation 1 must be used to interchange rows so that the pivot is nonzero.

Each pivoting iteration produces a 1 at the pivot position and zeros elsewhere in that column. The 1 is found by dividing the corresponding row by the pivot using operation 2, and the zeros are computed by operation 3 on the remaining rows.

THEOREM 1-13. *The result of any number of pivoting iterations on the augmented matrix of a system of linear equations is an equivalent system.*

PROOF: Pivoting uses only the three elementary row operations of Gauss-Jordan elimination. Thus, the resulting system has the same solution vectors, and is equivalent to the original system, by definition 19. ■

PROBLEMS

1-1. Given vectors $\mathbf{X} = (\ 2, 3, -1\ )$, $\mathbf{Y} = (-1, -2, 3\ )$, and $\mathbf{Z} = (4, -2, 5\ )$ find:
    (a) $(\ 2\mathbf{X} - \mathbf{Y}\ ) - 2\mathbf{Z}$
    (b) $3\mathbf{X} + (\ 4\mathbf{Y} - \mathbf{Z}\ )$

1-2. Using the vectors given in problem 1, illustrate the following laws of vectors:
    (a) Associative: $\mathbf{X} + (\ \mathbf{Y} + \mathbf{Z}\ ) = (\ \mathbf{X} + \mathbf{Y}\ ) + \mathbf{Z}$
    (b) Commutative: $\mathbf{X} + \mathbf{Y} = \mathbf{Y} + \mathbf{X}$
    (c) Distributive: $\alpha(\ \mathbf{X} + \mathbf{Y}\ ) = \alpha\mathbf{X} + \alpha\mathbf{Y}$ for any real number α.

1-3. Write a proof for each of the laws given in problem 2 for vectors in space R^n.

1-4. If $f(\ x\ )$ and $g(\ x\ )$ are real valued functions over $-1 \leq x \leq 1$ and $(\ f + g\ )x = f(\ x\ ) + g(\ x\ )$, $(\ \alpha f\ )x = \alpha f(\ x\ )$, show that the set of all such functions is a vector space.

1-5. Show that all vectors in the xz plane form a subspace of R^3.

1-6. Show that the origin is a subspace of R^3.

1-7. Show that the set of all functions $f(\ x\ )$ such that $f(\ 0\ ) = 0$ over $-1 \leq x \leq 1$ is a subspace of the space given in problem 4.

1-8. Show that the following transformation from R^3 to R^2 is linear:
$T(\ x_1, x_2, x_3\ ) = (\ x_1 - 3x_2, 2x_1 - x_2 + 4x_3\ )$

1-9. Show that the following transformation from R^3 to R^3 is linear:
$T(\ x_1, x_2, x_3\ ) = (\ x_1 + x_2 + x_3, x_1 - 2x_2 - 3x_3, 2x_1 - x_3\ )$

1-10. Find the following dot products using the vectors given in problem 1:
    (a) $\mathbf{X} \cdot \mathbf{Y}$
    (b) $\mathbf{X} \cdot \mathbf{Z}$
    (c) $\mathbf{Y} \cdot \mathbf{Z}$

1-11. For $\mathbf{X}$ and $\mathbf{Y}$ any vectors in R^n show that $\mathbf{X} \cdot \mathbf{Y} = \mathbf{Y} \cdot \mathbf{X}$

1-12. Using the vectors given in problem 1, show $\mathbf{X} \cdot (\mathbf{Y} + \mathbf{Z}) = \mathbf{X} \cdot \mathbf{Y} + \mathbf{X} \cdot \mathbf{Z}$

1-13. Show the distributive law given in problem 12 is true for all vectors in space R^n.

1-14. Show the following transformation is not linear: $T(x_1, x_2) = (2x_1 - 3, x_1 - x_2 + 4)$

1-15. Show that $T(x_1, x_2) = (x_1^2, 3x_2)$ is not linear.

1-16. Show that the projection of all vectors in R^3 onto the xy plane is a linear transformation.

1-17. If $T(x_1, x_2) = (x_1 + 2x_2, 3x_1 - 4x_2)$, find the matrix of the linear transformation.

1-18. Find the matrix $\mathbf{A}$ of the transformation in problem 9.

1-19. Describe the linear transformation corresponding to matrix

$$\mathbf{A} = \begin{pmatrix} 1 & 2 \\ 3 & 4 \\ 5 & 6 \end{pmatrix}$$

1-20. Determine the linear transformation corresponding to matrix

$$\begin{pmatrix} 3 & -4 & 2 & 6 \\ 2 & 1 & -5 & -7 \end{pmatrix}$$

1-21. Find the matrix $\mathbf{A}$ of the linear transformation in problem 8.

1-22. Compute

$$\begin{pmatrix} 3 & -2 & 5 & 4 \\ 1 & -6 & -7 & 8 \end{pmatrix} + \begin{pmatrix} 4 & -3 & 2 & -1 \\ 7 & -3 & 5 & -4 \end{pmatrix}$$

1-23. Find

$$\begin{pmatrix} a & b \\ c & d \end{pmatrix} + \begin{pmatrix} e & g \\ f & h \end{pmatrix}$$

1-24. Compute

$$5 \begin{pmatrix} 1 & 2 \\ 3 & -4 \\ -5 & 6 \end{pmatrix} -3 \begin{pmatrix} 2 & -4 \\ 6 & 8 \\ 7 & -9 \end{pmatrix}$$

1-25. Compute the products:

(a) $\begin{pmatrix} 1 & -2 & -3 \\ 4 & 5 & -6 \end{pmatrix} \begin{pmatrix} 2 & 0 \\ -3 & 5 \\ 4 & -1 \end{pmatrix}$

(b) $\begin{pmatrix} 2 & 0 \\ -3 & 5 \\ 4 & -1 \end{pmatrix} \begin{pmatrix} 1 & -2 & -3 \\ 4 & 5 & -6 \end{pmatrix}$

1-26. If $\mathbf{A} = \begin{pmatrix} 1 & 2 \\ 3 & 4 \end{pmatrix}$, $\mathbf{B} = \begin{pmatrix} -2 & 4 \\ 3 & -5 \end{pmatrix}$, $\mathbf{C} = \begin{pmatrix} 2 & 0 \\ -1 & 4 \end{pmatrix}$

verify the associative and distributive laws:
$\mathbf{A(BC)} = \mathbf{(AB)C}$, $\mathbf{A(B + C)} = \mathbf{AB} + \mathbf{AC}$

1-27. If $T(x_1, x_2) = 2x_1 - 3x_2, 4x_2, -x_1 + 3x_2 = \mathbf{Y}$ and $F(y_1, y_2, y_3) = 2y_1 - y_2 + y_3$, $y_1 - 3y_2, 3y_1 - 2y_2 - y_3 = \mathbf{Z}$ find the mapping $F(T(\mathbf{X})) = \mathbf{Z}$ and its matrix $\mathbf{C}$. Determine the matrix $\mathbf{A}$ of mapping F and matrix $\mathbf{B}$ of mapping T. Does $\mathbf{AB} = \mathbf{C}$?

1-28. Show $\mathbf{IA} = \mathbf{AI} = \mathbf{A}$ where $\mathbf{I}$ is the identity matrix in R^3 and

$$\mathbf{A} = \begin{pmatrix} a & b & c \\ d & e & f \\ g & h & i \end{pmatrix}$$

1-29. Given $T(x_1, x_2) = x_1 - 3x_2, 2x_1 + 4x_2 = \mathbf{Y}$, find the inverse transformation $T^{-1}(\mathbf{Y}) = \mathbf{X}$ by solving the system of linear equations for the column vectors $\mathbf{I}_j$ in terms of the unit vectors $\mathbf{i}_k$. Find the matrices $\mathbf{A}$ and $\mathbf{A}^{-1}$ corresponding to T and T^{-1}. Check that $\mathbf{AA}^{-1} = \mathbf{I}$

1-30. Find the inverse of $\mathbf{A} = \begin{pmatrix} 0 & 1 & 1 \\ 1 & 0 & -1 \\ 0 & 0 & 1 \end{pmatrix}$

Can a nonidentity matrix be its own inverse?

1-31. Determine the inverse of $\mathbf{A} = \begin{pmatrix} 1 & 1 \\ -1 & 1 \end{pmatrix}$ and check.

1-32. Find the inverse of the $\mathbf{A}^{-1}$ found in problem 31. Does $(\mathbf{A}^{-1})^{-1} = \mathbf{A}$?

1-33. For $\mathbf{A} = \begin{pmatrix} 1 & 1 \\ -1 & 1 \end{pmatrix}$ and $\mathbf{B} = \begin{pmatrix} 1 & 2 \\ 3 & 4 \end{pmatrix}$, verify that $(\mathbf{AB})^{-1} = \mathbf{B}^{-1}\mathbf{A}^{-1}$.

1-34. Reduce the following matrix to echelon form and give the solution to the corresponding system of linear equations.

$$\begin{pmatrix} 1 & 0 & 2 & 1 & 3 \\ 2 & 1 & 0 & 2 & 0 \\ 0 & 2 & 1 & 0 & -3 \\ 1 & 2 & 0 & 3 & 1 \end{pmatrix}$$

1-35. Is the system of linear equations corresponding to the following augmented matrix inconsistent?

$$\begin{pmatrix} 2 & -8 & 4 & 3 \\ 1 & 2 & 0 & -2 \\ -3 & 12 & -6 & 1 \end{pmatrix}$$

1-36. Express the solution set corresponding to the following augmented matrix.

$$\begin{pmatrix} 1 & 2 & 3 & 4 & 3 \\ 2 & 2 & 0 & 3 & 4 \\ 2 & 4 & 6 & 8 & 6 \end{pmatrix}$$

1-37. Solve by Gauss-Jordan elimination:

$$\begin{pmatrix} 4x_2 - 7x_3 = 13 \\ 3x_1 + 5x_3 = 0 \\ 6x_1 + 11x_2 = 8 \end{pmatrix}$$

1-38. Find the rank of the following matrices:

(a) $\begin{pmatrix} 1 & 3 \\ -2 & -4 \\ 5 & 0 \end{pmatrix}$

(b) $\begin{pmatrix} 1 & 2 & 0 & 4 \\ 1 & -1 & 3 & -2 \\ 2 & 4 & -2 & 0 \end{pmatrix}$

(c) $\begin{pmatrix} 4 & 3 & 1 & 2 & 5 \\ 2 & -1 & 1 & 0 & -1 \\ 3 & -4 & 2 & -1 & -5 \\ 11 & 2 & 4 & 3 & 5 \end{pmatrix}$

1-39. Decide whether the following systems have no solution, one solution, or many solutions:

(a) $x_1 + 2x_2 = 3$
  $3x_1 + 4x_2 = 5$
  $4x_1 + 6x_2 = 8$

(b) $x_1 + 2x_2 \qquad = 4$
  $x_1 - x_2 + 3x_3 = -2$
  $2x_1 + 4x_2 - 2x_3 = 0$

(c) $2x_1 + 3x_2 - 4x_3 = 1$
  $3x_1 - 2x_2 \qquad = 2$
  $x_1 + 8x_2 - 8x_3 = 0$

(d) $x_1 \qquad - x_3 = 1$
  $-2x_1 + x_2 \qquad = 2$
  $-6x_1 + 5x_2 - 4x_3 = -1$

1-40. Find an elementary matrix from $\mathbf{I}^3$ that will take 2 times the second row and subtract it from 3 times the third row. As a check, multiply it times the matrix

$$\begin{pmatrix} a & b & c \\ d & e & f \\ g & h & i \end{pmatrix}$$

1-41. Find an elementary matrix from $\mathbf{I}^3$ that will change the order of the rows to 3, 1, 2 and check by an appropriate product.

1-42. Find the inverse of matrix $\mathbf{A} = \begin{pmatrix} 4 & 8 \\ 3 & 2 \end{pmatrix}$ and check by computing $\mathbf{AA}^{-1}$.

1-43. Compute the inverse of the following matrix by pivoting.

$$\begin{pmatrix} 2 & 3 & 4 \\ 4 & -3 & 2 \\ 0 & 1 & 2 \end{pmatrix}$$

1-44. Determine if the following matrices are singular:

(a) $\begin{pmatrix} 4 & -8 \\ 3 & -6 \end{pmatrix}$

(b) $\begin{pmatrix} 1 & 0 & -1 \\ -2 & 1 & 0 \\ -6 & 5 & -4 \end{pmatrix}$

1-45. Find the inverse of the linear transformation
$T(x_1, x_2, x_3) = x_1 + 2x_2 + 3x_3, 2x_1 - x_2 + x_3, 3x_1 + 2x_2 - x_3$
by pivoting its matrix.

1-46. Solve:
$$\begin{aligned}
x_1 - x_2 - 2x_3 + x_4 &= -4 \\
x_1 - 2x_2 - 2x_3 + 2x_4 &= -5 \\
-x_1 + 2x_2 + 3x_3 - 2x_4 &= 6 \\
2x_1 - x_2 - 4x_3 &= -9
\end{aligned}$$

1-47. Solve and express the solutions in vector notation:
$$\begin{aligned}
x_1 + 2x_2 + x_3 - x_4 &= 2 \\
x_1 - x_2 - x_3 + 2x_4 &= 3 \\
2x_1 + 3x_2 + x_3 - x_4 &= -4 \\
3x_1 + 2x_2 + x_4 &= -1
\end{aligned}$$

1-48. Solve and express the solutions in vector notation:
$$\begin{aligned}
x_1 - x_2 + x_3 + x_5 &= 1 \\
2x_1 + 2x_3 + x_4 &= 3 \\
-3x_1 + x_2 - 3x_3 - x_4 + 2x_5 &= -2 \\
x_1 + 3x_2 + x_3 + 2x_4 + 3x_5 &= 7
\end{aligned}$$

1-49. Show vector $(3, 1)$ is a linear combination of vectors $(4, 3)$ and $(2, 4)$, that is, find the appropriate constants in definition 6.

1-50. Does the set of vectors $(1, 0, 1), (1, 2, 3), (3, 2, 1)$ span space R^3? If so, find the appropriate constants in terms of an arbitrary vector (y_1, y_2, y_3).

1-51. Does the set of vectors $(-1, 0, 1)$, $(1, 2, 3)$, $(3, 2, 1)$ span space R^3? Explain.

1-52. Vectors $(1, 2)$ and $(-2, b)$ are to span R^2. What are the only values that are not possible for the number b?

1-53. Are the vectors $(-1, 0, 1)$, $(1, 2, 3)$, $(3, 2, 1)$ linearly dependent? If so, find the appropriate constants so that a linear combination of these vectors equals the zero vector.

1-54. From the solution to problem 1-53 find vector $(3, 2, 1)$ as a linear combination of $(-1, 0, 1)$ and $(1, 2, 3)$. Check the result directly by using definition 6.

1-55. Are $(1, 0, 2, 3)$ and $(4, 5, 6, 0)$ linearly independent in space R^4?

1-56. Test the set of vectors $\{(1, 2), (3, 4)\}$ to see if it is a basis for R^2. Show whether it spans R^2 and is linearly independent in R^2.

1-57. Is the set of vectors $\{(-1, 0, 1), (1, 2, 3), (3, 2, 1)\}$ a basis for R^3?

1-58. Find the coordinates of $(1, 2, 3)$ in R^3 with respect to the basis $\{(0, 1, 1), (1, 0, 1), (1, 1, 0)\}$.

1-59. Suppose that a set of n vectors in space R^n is linearly independent, that is, the set is a basis for R^n. If an n by n matrix $\mathbf{A}$ is made up of the vectors in this set as column vectors, then what is the rank of $\mathbf{A}$? Prove your result.

1-60. If n by n matrix $\mathbf{A}$ is nonsingular, can the column vectors of $\mathbf{A}$ be used as a basis for space R^n?

CHAPTER 2

Graphical Methods

2.1 INTRODUCTION

Linear programs can solve economic problems of industry and government that could not be handled otherwise due to the enormous expense or time involved. For example, the techniques of linear programming have found the cheapest, fastest, or most profitable way for large enterprises to function. They tell management how to move supplies, allocate personnel, or utilize resources in the most efficient manner. The decision makers use linear programs to determine plant location and production schedules. In general, these programs have helped to find optimum choices in numerous competitive situations.

A part of the popularity of linear programming is due to our increased use of computers. The simultaneous development of fast computers has provided the necessary equipment to handle large problems that previously could not be solved because of lack of time or money. All that was necessary after the technical advance in electronic computing was the discovery and development of a theory that could be programmed on a digital computer. George B. Dantzig proposed such a theory in 1947 while doing research for the U.S. department of the Air Force. He originated the *Simplex Method* for solving linear problems. The first run of linear program by Dantzig's method occurred in 1952 on the computer of the National Bureau of Standards in Washington, D.C. Since then the method has been programmed for practically all computers, large or small. Computer routines that may be adapted to any machine are found in Chapter 5.

In this chapter the terminology of linear programming will be explained and a graphical analysis used to solve two-dimensional cases. These simple graphs will provide a background of what to expect in the general n-dimensional problem.

2.2 THE LANGUAGE OF LINEAR PROGRAMMING

The problems we will deal with are called *linear programs* because they are composed of linear expressions. As the word *linear* suggests, all the variables occurring in a linear program are first degree. A linear program is characterized by a first degree function of n variables that is to be maximized or minimized, subject to a set of m first degree constraints. A *constraint* is an equality or weak inequality[1] representing a restriction on the variables of a program.

Since the variables in a linear program usually stand for physical objects, they are most often required to be nonnegative. Thus a typical linear program is stated as:

Find $x_j \geq 0$, $j = 1, 2, \ldots, n$, that will maximize (or minimize)

$$\sum_{j=1}^{n} c_j x_j \quad \text{such that} \tag{1}$$

$$\sum_{j=1}^{n} a_{ij} x_j \leq b_i \text{ for } i = 1, 2, \ldots, m. \tag{2}$$

The x_j's are the variables, while a_{ij} and b_i in the m linear constraints of (2), as well as the c_j in the linear function of (1), are given constants for all subscripts. Geometrically the x_j's are the coordinates of a point (a vector) in n-dimensional space. R^n.

The same linear program may be stated in matrix form as: Find vector **X**, all of whose components are nonnegative, that will maximize (or minimize)

CX   such that

AX $\leq$ **B**.

A is an m by n coefficient matrix, **X** is a n by 1 column vector of unknowns, **B** is an m by 1 column vector of constants, and **C** is a 1 by n row vector of constants.

DEFINITION 1.   The coordinates of a point that satisfies all n linear constrains in equation (2) constitute a *solution*.

DEFINITION 2.   If the coordinates of a solution point are all nonnegative, then the solution is called *feasible*.

---

[1] A *weak inequality* may possibly be satisfied by equality.

DEFINITION 3.   If one or more of the coordinates of a solution point are negative, then the solution is an *infeasible solution*. An *infeasible linear program* is a problem in which all solution points are infeasible or a problem that has no solution points.

DEFINITION 4.   The set of all feasible points is known as the *feasible region* of the problem.

DEFINITION 5.   If among the feasible solutions there is one for which the linear function in (1) has its maximum (minimum) value, then this is called an *optimal feasible solution*.

DEFINITION 6.   The linear function in (1) is the *objective function* and its value at a point in the feasible region is referred to as the *objective value*.

From definitions 5 and 6 it is clear that an optimal feasible solution is a feasible solution for which the objective value is maximum (minimum).

Later we will show how the above format may include constraints that are equations, and variables that are unrestricted in sign.

2.3   GRAPHICAL ANALYSIS: MAXIMIZING

Consider the following linear program.

Example 1.   Fretmor is worried about his two toughest subjects, calculus and chemistry. After taking into account the rest of his subjects, he realizes he must get at least 3 grade points out of calculus and chemistry combined to maintain his class standing. Of course, if he could pull an *A* in either one, he would get the maximum of 4 grade points. Experience has shown Fretmor that for each grade point in calculus he must study 10 minutes a day and for each grade point in chemistry he must study 15 minutes a day. Fretmor is unwilling to budget more than a total of 70 minutes a day for these two subjects. Under these conditions what is the maximum number of grade points he can earn and how should his time be divided between calculus and chemistry?

SOLUTION.   First choose the unknowns. Let x equal the number of grade points he receives in calculus and let y equal the number of grade points he receives in chemistry. The condition that he must get at least 3 grade points imposes the following constraint on the unknowns:

$$x + y \geq 3. \tag{1}$$

This is a weak inequality and allows the sum of the grade points to be equal to 3 or greater than 3. The fact that 8 is the greatest number of grade points attainable imposes an upper bound constraint:

$$x + y \leq 8. \tag{2}$$

Inequality (2) will be called a *type I inequality* ($\leq$) while inequality (1) will be called a *type II inequality* ($\geq$). Both type I and type II will allow the case of equality to be true.

Further constraints on the unknowns are

$$x \leq 4, \qquad y \leq 4, \tag{3}$$

since 4 is the greatest number of grade points possible in one subject, and

$$x \geq 0, \qquad y \geq 0, \tag{4}$$

since zero is the least number of grade points possible. Constraints (4) are called nonnegativity constraints. A final constraint imposed by the time requirement is

$$10x + 15y \leq 70. \tag{5}$$

This inequality may be simplified by dividing both sides by 5,

$$2x + 3y \leq 14.$$

The object of the problem is to maximize the number of grade points, so we find the largest possible number M such that

$$M = x + y. \tag{6}$$

The linear program is formulated as: Find x and y such that $M = x + y$ is as large as possible and such that

$$\begin{aligned}
x + y &\geq 3 \\
x + y &\leq 8 \\
x &\leq 4 \\
y &\leq 4 \\
2x + 3y &\leq 14,
\end{aligned} \tag{7}$$

where x and y are nonnegative real numbers.

To solve the problem graphically, graph the region of the Cartesian plane corresponding to each of the constraints in (7) and (4). First draw the straight line $x + y = 3$, a line of slope -1 through (3, 0). A linear inequality in two variables is a half plane. The equality or straight line is called the *boundary* of the half plane. The half plane along with its boundary is called a *closed half plane*. We must decide on which side of the line $x + y = 3$ our half plane lies. An easy way is to solve the inequality for y:

$$y > 3 - x.$$

For fixed x the ordinates satisfying this inequality are larger than the corresponding ordinate on the line, and thus the inequality is satisfied for all points above the line. The region is shown in Figure 2–1.

Graph the line $x + y = 8$ in the same Cartesian plane and shade the closed half plane $y \leq 8 - x$, which lies on or below the line. Since we must satisfy both constraints simultaneously, we want the set intersection of these two closed half planes, or the strip between the two parallel lines, as shown in Figure 2–2.

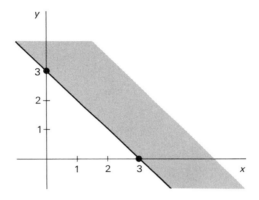

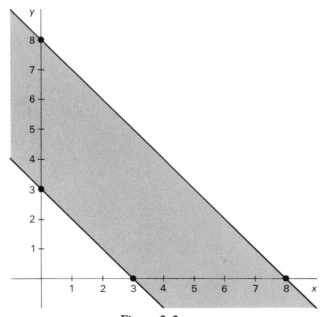

Figure 2-2

We are restricted to the first quadrant portion of this strip because x and y are nonnegative. Now intersect the first quadrant strip with the closed half planes $x \le 4$ and $y \le 4$, as shown in Figure 2–3. Note from this graph that the constraint $x + y \le 8$ is redundant. We get the same shaded region if $x + y \le 8$ is omitted. This frequently happens in a linear program but seldom causes any trouble.

Finally intersect this region with the closed half plane $y \le (14 - 2x)/3$ which is on or below the line $2x + 3y = 14$ of slope $-\frac{2}{3}$. The shaded region in Figure 2–4 represents the feasible region of the linear program.

In our problem the objective function is $f(x, y) = x + y$. If we consider M as a parameter and graph $M = x + y$ for various values of M, we have a family of parallel lines with slope -1. Some of these lines will intersect the feasible region and contain many feasible solutions while others will miss and contain no feasible solution. We wish to find the line of this family that intersects the

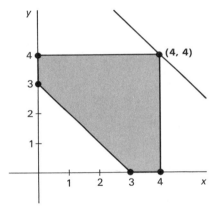

Figure 2-3

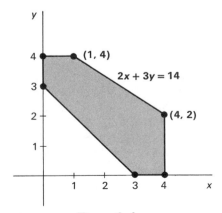

Figure 2-4

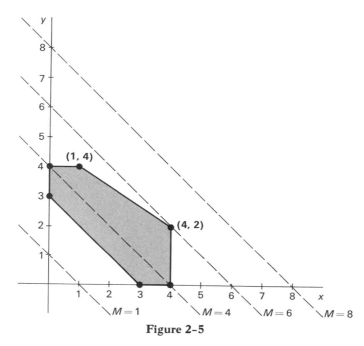

Figure 2-5

feasible region and is farthest out from the origin. The farther from the origin the greater will be the value of M.

From the graph in Figure 2–5, we see that $M = 6$ does the job. It intersects the feasible region at the vertex $(4, 2)$. Since $(4, 2)$ is the only feasible point on $M = 6$, this solution is unique. Thus, Fretmor can achieve a maximum of 6 grade points, 4 for an A in calculus and 2 for a C in chemistry, by studying 40 minutes a day on calculus and 30 minutes a day on chemistry, using up his top allotted time of 70 minutes per day.   •

2.4   GRAPHICAL ANALYSIS: MINIMIZING

Consider another example of a linear program, a case in which the objective function is to be minimized.

Example 2.   Candidate Wynnles is trying to hold a strict budget and decide how many personal appearances versus TV appearances he should make in the state presidential primary. His committee has figured that the total expense of each personal appearance at a campaign rally is $15,000 while the cost of each TV speech is $12,000. Mr. Wynnles has an efficiency expert who has estimated that each personal appearance rally will net 30,000 votes while each TV appearance will net 40,000 votes. Candidate Wynnles knows that he needs at least 240,000 votes to swing the state. Each of his personal appearance trips uses up 2 days while a live TV run takes only 1 day. He will spend at least 10 days in this state. For each personal appearance the party will provide him with 50 precinct workers but for a TV appearance the party can guarantee only 30 precinct workers. The local party boss claims that the winner must have at least 290 precinct workers. What is the minimum cost at which candidate Wynnles can carry this state?

SOLUTION.   Let x equal the number of personal appearances and y equal the number of TV appearances. The constraint on the number of votes to win is a type II inequality:

$$30,000x + 40,000y \geq 240,000.$$

Likewise we have a lower bound constraint on the number of days to be spent:

$$2x + y \geq 10.$$

The final constraint on the number of precinct workers is also of type II:

$$50x + 30y \geq 290.$$

The objective function is the total cost of the campaign:

$$\$15,000x + \$12,000y.$$

After the inequalities and objective function are simplified by dividing each through by a positive constant, the problem may be restated as follows: Find nonnegative numbers x and y such that

$$\begin{aligned} 3x + 4y &\geq 24 \\ 2x + y &\geq 10 \\ 5x + 3y &\geq 29, \end{aligned} \qquad (1)$$

where $m = 5x + 4y$ is to be a minimum.

As in the first example, graph the closed half plane corresponding to each constraint and find the intersection of these half planes in the first quadrant. See Figure 2–6. Note in each case the half plane lies above its boundary line. The feasible region, as indicated by the shading in Figure 2–6, is infinite in extent. It includes all of the first quadrant above the boundary lines. A maximum in this case would be impossible to find, but we are looking for a minimum. Let us consider m a parameter in $m = 5x + 4y$ and draw the line for various values of m. We seek a line in this family of lines that intersects the feasible region and at the same time is as close as possible to the origin. Note in Figure 2–7 that the appropriate line is $m = 32$, and that the optimal feasible solution occurs at vertex (4, 3). Each of the vertices was found by solving the corresponding pair of linear equations for their intersection. The optimal point (4, 3) tells us that candidate Wynnles should make four personal appearances and also three TV

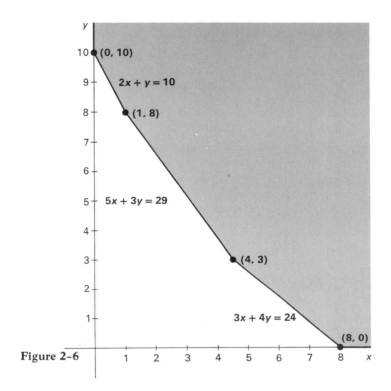

Figure 2-6

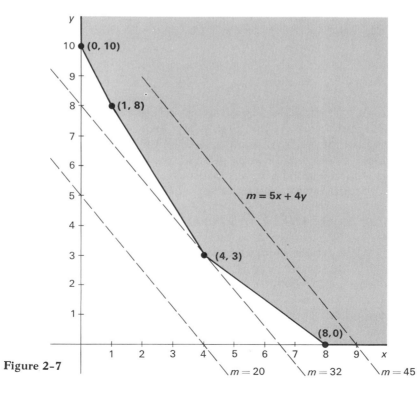

Figure 2-7

speeches to win the primary. By converting to the original units, we compute his minimum cost to be

$$\$15,000(\ 4\ ) + \$12,000(\ 3\ ) = \$96,000.$$

We may also check the original constraints to see that the optimal solution produces the required 290 precinct workers and the minimum of 240,000 votes. However, the remaining constraint shows that an additional day of Mr. Wynnles' time is required above the minimum 10 days. This additional day is known as the *slack* in that problem constraint. It is the quantity that must be added to or subtracted from an inequality to force it to equality. •

Once again our solution is unique, but this is not necessarily the case as the next example will show.

2.5 MULTIPLE SOLUTIONS

Example 3.   Suppose in example 2, the constraints remain the same but the objective function is changed by charging $12,000 for a personal appearance and $16,000 for TV time. Now what is the minimum cost for candidate Wynnles to win?

The problem may be stated as: Find nonnegative numbers x and y such that

$$3x + 4y \geq 24$$

$$2x + \ y \geq 10$$

$$5x + 3y \geq 29,$$

where $m = 3x + 4y$ is a minimum.

SOLUTION.   The feasible region is the same as before but note in Figure 2–8 that one member of the family $m = 3x + 4y$ passes through a boundary line. This means that many solutions exist. Not only will vertex (4, 3) do but also the vertex (8, 0). Actually any point on the line segment between these two vertices is an optimal feasible solution.

If we convert back to cost, the vertex (4, 3) corresponds to

$$\$12,000(\ 4\ ) + \$16,000(\ 3\ ) = \$96,000.$$

Vertex (8, 0) gives the same minimum cost,

$$\$12,000(\ 8\ ) + \$16,000(\ 0\ ) = \$96,000.$$

Any other point on this boundary gives the same minimum cost. For example (6, ³⁄₂) gives

$$\$12,000(\ 6\ ) + \$16,000(\ ³⁄₂\ ) = \$96,000.$$

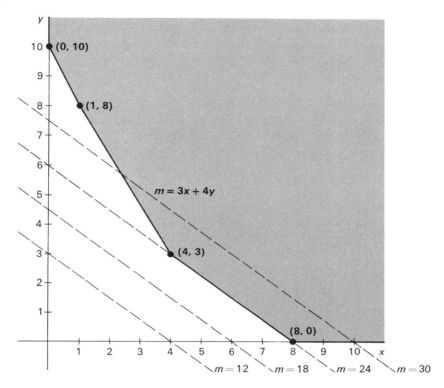

Figure 2-8

Of course from a practical viewpoint, Mr. Wynnles can't make a fractional appearance, so he would only be interested in integral answers. He could now choose the additional solution of eight personal appearances and no TV appearances at the same cost. Both integral solutions produce the same number of votes, but (8, 0) would require 16 days of his time.   •

2.6 SUMMARY

Before proceeding to the theory, we may draw some inferences from our solutions in two dimensions. For the optimal solution to be *unique*, it is necessary that the graph of the objective function touch the feasible region at a single vertex. The coordinates of this vertex make up the optimal feasible solution. It is possible for *many* optimal solutions to exist along a boundary of the feasible region. It is not possible to find an optimal solution strictly inside the feasible region. If the graph of the objective function passed through such an interior point, then the graph could be moved either in the direction of increasing M or decreasing m while still intersecting the feasible region. Finally if the feasible region is unbounded, it might be possible to move the objective graph arbitrarily far from the origin while still passing through feasible points. In this case the corresponding objective value could be *unbounded.*

A final possibility is that there are no feasible points that satisfy all of the given constraints. A program with an empty feasible region has no feasible solution, and the program is called *infeasible* or *inconsistent*.

PROBLEMS

For the first three problems, draw the graph of the region defined by the given linear constraints. For the sketched region find the maximum and the minimum of the objective function. Are all of the given constraints essential or may the number of constraints be reduced without changing the feasible region?

2–1. $x \geq 0, y \geq 0$
$2x + 3y \geq 11$
$3x + y \geq 6$
$4x - 5y \geq -30$
$5x + 2y \leq 45$
$4x - 3y \leq 13$
objective function $x + 3y$

2–2. $x \geq 0, y \geq 0$
$x + 2y \geq 4$
$x + 2y \leq 10$
$y - x \leq 5$
$x - y \leq 5$
objective function $2x + y$

2–3. $x \geq 0, y \geq 0$
$x + 2y \geq 4$
$x + 4y \leq 20$
$y \leq x + 5$
$x \leq y + 5$
objective function $x + y$

For the following problems, use graphical methods to find the optimum solutions if such exist. Note whether or not the solutions are unique. In the case of multiple solutions, give several answers.

2–4. $x \geq 0, y \geq 0$
$2x + y \geq 11$
$x + 2y \geq 10$
$3x + y \geq 13$
$x + 4y \geq 12$
Minimize $13x + 5y$.

2–5. Under the same constraints given in problem 2–4, minimize $17x + 34y$. Does this function have a maximum over the feasible region?

2-6. $x \geq 0, y \geq 0$
$3x + 5y \leq 55$
$4x - 5y + 20 \geq 0$
$5x + 2y \leq 60$
Maximize $7x + 4y$; minimize $7x + 4y$.

2-7. Under the same constraints given in problem 2–6, minimize $9x - 2y$. Does this function have a maximum? If so find it.

2-8. $x \geq 0, y \geq 0$
$3x - 7y \leq 9$
$3x - 7y \geq -14$
Maximize $6x - 21y$; is there a minimum value for this function over the feasible region?

2-9. $x \geq 0, y \geq 0$
$7x + 6y \geq 70$
$2x + 5y \leq 30$
$x - 2y \leq -10$
Maximize $ax + by$ for any real constants a and b; are there any feasible points?

2-10. $x \geq 0, y \geq 0$
$5x + 3y \geq 21$
$3x - 5y \geq -35$
$5x - 4y \leq 35$
$5x + 6y \leq 85$
$x + 2y \geq 7$
Maximize $15x + 18y$.

2-11. For the same constraints given in problem 2–10, minimize $15x + 18y$.

2-12. For the constraints given in problem 2–10, find both the maximum and the minimum of $3x + 2y$.

2-13. Mrs. Coffman was told to supplement her daily diet with at least 6000 USP units of vitamin A, at least 195 mg. of vitamin C, and at least 600 USP units of vitamin D. Mrs. Coffman finds that her local drug store carries blue vitamin pills at 5 cents each and red vitamin pills at 4 cents each. Upon reading the labels she sees that the blue pills contain 3000 USP units of A, 45 mg. of C and 75 USP units of D, while the red pills contain 1000 USP units of A, 50 mg. of C, and 200 USP units of D. What combination of vitamin pills should Mrs. Coffman buy to obtain the least cost? What is the least cost per day?

2-14. Seeall Manufacturing Company makes color TV sets. They produce a bargain set that sells for $100 profit and a deluxe set that sells for $150 profit. Along the set assembly line the bargain set requires 3 hours while the deluxe set takes 5 hours. The cabinet shop spends one hour on the cabinet for the bargain set and 3 hours on the one for the deluxe set. Both sets use 2 hours of time for testing and packing. On a particular production run the Seeall company has available 3900 man-hours on the assembly line, 2100 man-hours in the cabinet shop, and 2200 man-hours in the testing and packing department. How many sets of each type should they produce and what is their maximum profit?

2–15. Mr. Wise is head of the sporting goods department of the Eure Department Store. He decides to cut his inventory of spinning rods and reels by having a gigantic sale on 80 rods and 75 reels. Mr. Wise figures he will sell the $15 rods at a loss for a "come on" while selling the $32 spinning reels at a profit. He knows that most customers buying a rod will also pick up a new reel to match. The problem is to determine how much of a loss he can take on rods and how much of a mark up he needs on reels to sell the lot at a maximum profit. Mr. Wise knows from experience that to sell all the rods and reels he must attract at least 400 customers with his advertisement. For each dollar loss on a rod the ad will attract 400 customers, but each dollar mark up on reels will discourage 100 possible buyers from coming to the sale. The Eure store manager insists that a reel must sell at a profit at least as big as the loss on a rod. The mark up on a reel cannot exceed $6 and still be competitive. Finally Mr. Wise realizes that a rod and reel together should not sell for more than $49 to be sure that most go in sets. At what price does Mr. Wise advertise rods and what is his selling price on reels assuming that all are sold? What is the profit from this sale?

2–16. In problem 2–15, suppose only 65 rods and 70 reels actually sell. Then what should have been the selling prices and the profit?

2–17. The Neely Nut Company sells mixtures of shelled nuts. Currently the company has on hand 10,000 pounds of shelled Arizona pecans and 9000 pounds of shelled English walnuts. Under the label of *Nut Delight* they sell a cardboard box containing one pound of shelled Arizona pecans and 1½ pounds of shelled English walnuts. Another popular seller under their patented brand name of *Walcans*, is a plastic bag containing two pounds of shelled Arizona pecans and 1½ pounds of shelled English walnuts. The packaging equipment, which can only handle one brand at a time requires a half minute per box to fill, seal, and label the *Nut Delight*, while only 12 seconds per bag is required on the *Walcans*. The management is facing an impending employee strike in 37½ working hours. They wish to realize the maximum profit possible before the strike. If their profit is $1.60 on a box of *Nut Delight* and $1.50 on a bag of *Walcans*, then how should Neely Nut Company package its nuts?

2–18. If the strike is settled between the employees and management of Neely Nut Company, then how should the nuts be packaged in the previous problem to maximize profit?

2-19. The Steeping Tea Company is preparing two popular brands of tea, brand A and brand B. The quality of tea is determined by the position of the leaf on a twig of the tea bush. Orange pekoe is made from the first leaf after the bud at the tip of the twig. Pekoe is made from the second leaf, and souchong first is made from the third leaf. Steeping Tea has perfected its own blends. Brand A is a blend of 50% orange pekoe, 25% pekoe, and 25% souchong first. Brand B is a blend of 30% orange pekoe, 25% pekoe, and 45% souchong first. To meet back orders they will blend at least 550 pounds of orange pekoe at a cost of $4.00 per pound, at least 375 pounds of pekoe at a cost of $2.40 per pound, and at least 505 pounds of souchong first at a cost of $1.60 per pound. How should Steeping Tea Company blend this tea into brands A and B so as to minimize the total cost of the two brands? What is the minimum total cost?

2-20. The White Glove Catering Service has a problem in planning for the men's bowling league banquet. It has been decided that the main dish will be a choice between a seafood platter and prime roast of beef. The league secretary reports that at least 300 members will attend the banquet. The president says that at least 50 more of his bowlers are beef eaters than seafood eaters. The banquet chairman has requested that at least 90 seafood platters be prepared. The number of platters of prime roast of beef that the manager of White Glove will prepare can be no greater than 130 more than the number of seafood platters, because then seafood would be left over and wasted. If the actual cost of preparing the seafood platter is $6.75 and the cost of the prime roast of beef platter is $8.50, then what combination of the two would produce the minimum cost to White Glove?

2-21. If the cost of seafood increased so that a seafood platter cost $8.60, then what combination of dinners would yield the minimum cost to White Glove?

CHAPTER 3

Convexity

3.1 DEFINITION AND PROPERTIES OF CONVEX SETS

One of the reasons that we could draw the conclusions reached in section 2.6 is that the feasible regions of linear programs are *convex regions*.

DEFINITION 1. A *region* or set R is *convex* if and only if for every two points of R the line segment connecting those points lies entirely in R.

The definition assumes that set R contains at least two points. The special cases where R equals either the empty set or a set of only one point are also taken to be convex regions. A theorem that follows easily from definition 1 gives the following significant result.

THEOREM 3-1. *The intersection of any number of convex regions is convex.*

PROOF. Start with a given collection of convex regions and let R be their intersection. R consists of all points common to each region in the collection. Thus if P_1 and P_2 are points belonging to R, they belong to each region of the original collection. Since each given region is convex, the segment P_1P_2 belongs to each given region and therefore P_1P_2 belongs to R. Thus R satisfies definition 1. ∎

To make definition 1 workable, we need an algebraic expression for all the points on a line segment. This is most easily done by introducing a parameter, α, and writing the equations of the line segment in parametric form. Let $\mathbf{X}$ and $\mathbf{Y}$ be two arbitrary points in m-dimensional space with their coordinates given by x_i, $i = 1$ to m and y_i, $i = 1$ to m, respectively. Then the line segment $\mathbf{XY}$ is the set of all points $\mathbf{U}$ with coordinates satisfying the parametric equations

$$u_i = (1 - \alpha)x_i + \alpha y_i, \qquad i = 1, \ldots, m \qquad \text{where } 0 \le \alpha \le 1. \tag{1}$$

Any of the points $\mathbf{U}$ is called a *convex combination* of points $\mathbf{X}$ and $\mathbf{Y}$. The set of all convex combinations of $\mathbf{X}$ and $\mathbf{Y}$ for $0 \le \alpha \le 1$ is precisely the line segment $\mathbf{XY}$, which is convex.

For the remainder of this chapter we will concentrate on the two-dimensional plane where the relationships are easy to see and draw. The ideas must be extended to higher dimensional space to apply to general linear programs. In two dimensions if $\mathbf{X}(x_1, x_2)$ and $\mathbf{Y}(y_1, y_2)$ are points in a plane, then from the parametric equations (1) the line segment $\mathbf{XY}$ is the set of points $\mathbf{U}(u_1, u_2)$ satisfying the pair of parametric equations

$$\begin{aligned}
u_1 &= (1 - \alpha)x_1 + \alpha y_1 \\
u_2 &= (1 - \alpha)x_2 + \alpha y_2, \qquad \text{for } 0 \le \alpha \le 1.
\end{aligned} \tag{2}$$

See Figure 3–1.

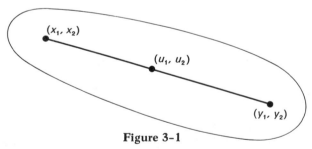

Figure 3–1

Let R be a plane convex region. The definition of convexity states that the set of points satisfying parametric equations (2) belongs to R whenever (x_1, x_2) and (y_1, y_2) belong to R. Note that the values 0 and 1 of the parameter α produce the end points $\mathbf{X}$ and $\mathbf{Y}$ respectively. The remaining real values of α between 0 and 1 produce points along $\mathbf{XY}$ at those proportionate distances from $\mathbf{X}$. For example $\alpha = \frac{1}{2}$ results in the familiar midpoint formulas, while $\alpha = \frac{1}{3}$ will place $\mathbf{U}$ one-third of the way from $\mathbf{X}$ to $\mathbf{Y}$. The line segment defined by equations (2) is called a *closed segment* because it includes its end points $\mathbf{X}$ and $\mathbf{Y}$. An *open segment* would omit the end points. If α is an unrestricted real number, then equations (2) would represent the entire line defined by points $\mathbf{X}$ and $\mathbf{Y}$.

We saw in Chapter 2 that a linear constraint $ax_1 + bx_2 \le c$ defined a closed half plane where the boundary line corresponds to equality in the constraint.

THEOREM 3-2. *The closed half plane $ax_1 + bx_2 \leq c$ is a convex region.*

PROOF.   Let $Y(\,y_1,\,y_2\,)$ and $Z(\,z_1,\,z_2\,)$ belong to the closed half plane $ax_1 + bx_2 \leq c$. Thus their coordinates satisfy

$$ay_1 + by_2 \leq c$$
$$az_1 + bz_2 \leq c.$$

Let $U\,(\,u_1,\,u_2\,)$ be any point on the line segment YZ. From the parametric equations of YZ we have

$$au_1 + bu_2 = a[\,(\,1-\alpha\,)y_1 + \alpha z_1\,] + b[\,(\,1-\alpha\,)y_2 + \alpha z_2\,]$$
$$= (\,1-\alpha\,)(\,ay_1 + by_2\,) + \alpha(\,az_1 + bz_2\,)$$
$$\leq (\,1-\alpha\,)c + \alpha c, \qquad \text{since both } \alpha \text{ and } 1-\alpha \text{ are nonnegative,}$$
$$\leq c.$$

This proves that all points on the line segment YZ are in $ax_1 + bx_2 \leq c$, and therefore the closed half plane is convex.   ■

An open half plane is also convex, but is not of interest in linear programming. In m-dimensional space, $m > 2$, the correspondent to a closed half plane is a closed half space. The closed half space may be proved to be a convex region by allowing the subscripts in theorem 3–2 to run from one to m.

Combining theorems 3–1 and 3–2 gives the following theorem.

THEOREM 3-3. *The intersection of any number of half planes is a convex region.*

DEFINITION 2.   A *polygonal convex region* is the intersection of a positive finite number of closed half planes.

The feasible region of a finite set of linear inequalities in two variables is the intersection of a finite number of closed half planes and is therefore a polygonal convex region. The feasible regions discussed in Chapter 2 were all polygonal convex regions.

DEFINITION 3.   A *convex polygon* is a polygonal convex region R such that any line through a point of R intersects R in a closed segment.

Since a closed segment has finite length, a convex polygon is a bounded region of the plane. The boundaries of a polygonal convex region are made up of either all or portions of the boundary lines of the intersecting planes. These boundaries are then either lines, rays, or closed segments. To see that each case may occur, refer to Figure 3–2.

In Figure 3–2(a) the boundaries of R are parallel lines. In (b) boundaries A and B are rays or half lines. All of the boundaries in (c) are line segments. Note that (a) and (b) typify general polygonal convex regions that

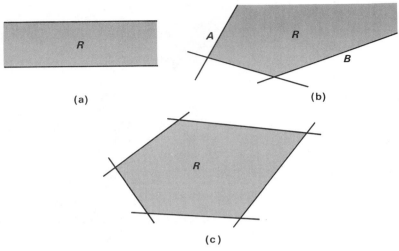

(a)

(b)

(c)

Figure 3-2

are unbounded or infinite in extent, and (c) illustrates a convex polygon that is necessarily a finite region of the plane.

The analog of a convex polygon in three dimensions is called a *convex polyhedron*. The faces of a convex polyhedron are convex polygons. All of the regular polyhedrons such as the tetrahedron and cube are convex. A characteristic of these figures is that they have corners, commonly called vertices or extreme points.

DEFINITION 4. A *vertex* or extreme point in a convex set R is a point of R that cannot be written in equations (1) as a convex combination of two other distinct points of R.

According to definition 4, vertices in a sense stand alone. They cannot be reached by any line segment connecting two points of R that are distinct from the vertex itself. In the case of convex polyhedra, the vertices will be where edges come together. An interesting convex region is the solid sphere because every point of its surface is an extreme point.

In two dimensions a vertex of a polygonal convex region is a point in the region where two boundary lines join. It is possible for more than two boundary lines to pass through a vertex, but then all except two of them will be redundant in defining the convex region. We might just as well say that for a polygonal convex region, a vertex is a point in the region on exactly two boundary lines.

3.2 LINEAR FUNCTIONS

In the plane with coordinates x_1 and x_2, let us examine linear functions of the form $f(x_1, x_2) = ax_1 + bx_2$ where a and b are real constants not both zero. If $\mathbf{Y}(y_1, y_2)$ and $\mathbf{Z}(z_1, z_2)$ are distinct points of the plane, then the *parametric equations* of the line through $\mathbf{Y}$ and $\mathbf{Z}$ are

$x_1 = (1 - \alpha)y_1 + \alpha z_1$

$x_2 = (1 - \alpha)y_2 + \alpha z_2,$ for all real numbers α.

Let $f(y_1, y_2) = m$ and $f(z_1, z_2) = M$. For any point $\mathbf{X}(x_1, x_2)$ on line $\mathbf{YZ}$ we have the function

$$f(x_1, x_2) = a[(1 - \alpha)y_1 + \alpha z_1] + b[(1 - \alpha)y_2 + \alpha z_2]$$
$$= (1 - \alpha)(ay_1 + by_2) + \alpha(az_1 + bz_2)$$
$$= (1 - \alpha)f(y_1, y_2) + \alpha f(z_1, z_2)$$
$$= (1 - \alpha)m + \alpha M$$
$$= m + \alpha(M - m). \tag{3}$$

Using this result it is easy to prove the following theorem.

THEOREM 3-4. *If f is linear and has the same value at two distinct points $\mathbf{Y}$ and $\mathbf{Z}$, then f remains constant along the line through $\mathbf{Y}$ and $\mathbf{Z}$. On the other hand if f has different values at $\mathbf{Y}$ and $\mathbf{Z}$, then at each point of the open segment $\mathbf{YZ}$, f will have a value strictly between its values at $\mathbf{Y}$ and $\mathbf{Z}$.*

PROOF.   In equation (3) let $m = M$ and then $f(x_1, x_2) = m$ is constant. If $m \neq M$, say $m < M$, then $M - m > 0$. Equation (3) may be considered a function of parameter α. Call it $f(\alpha)$.

$f(\alpha) = m + \alpha(M - m),$

$f(0) = m$ and $f(1) = M.$

For any two values of the parameter $\alpha_1 < \alpha_2,$

$$(M - m)\alpha_1 < (M - m)\alpha_2$$
$$m + (M - m)\alpha_1 < m + (M - m)\alpha_2$$
$$f(\alpha_1) < f(\alpha_2).$$

This means that f is a strictly increasing function of parameter α. That is, as α increases the functional values $f(\alpha)$ steadily increase. Now, let $\mathbf{P}(x_1, x_2)$ with parameter value $\bar{\alpha}$ be any point of the open segment $\mathbf{YZ}$ as in Figure 3-3. In the parametric equations of segment $\mathbf{YZ}$ parameter α is also strictly increasing so $0 < \bar{\alpha} < 1$. Then,

$f(0) < f(\bar{\alpha}) < f(1)$ or $m < f(x_1, x_2) < M.$ ■

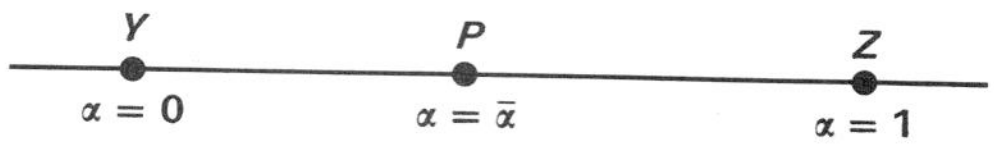

Figure 3-3

A conclusion of theorem 3–4 is that a linear function defined over a closed segment will assume its maximum and minimum at the end points of that segment.

3.3   THE EXTREME POINT THEOREM

We are particularly interested in the extreme points or vertices of a polygonal convex region. Fundamental in this development is the idea observed in Chapter 2 that a linear objective function attains its optimum over a convex region at a vertex of that region.

THEOREM 3-5.   *The maximum and minimum values of a linear function f defined over a convex polygon R exist[1] and are found at the vertices of R.*

PROOF.   Suppose $\mathbf{W}(w_1, w_2)$ is an arbitrary interior point of the given convex polygon R in a plane with coordinates (x_1, x_2), that is, $\mathbf{W}$ is a point of R not on the boundary of R. Let L be an arbitrary line through $\mathbf{W}$, a line which by definition 3 intersects R in a closed segment. Let $\mathbf{Y}(y_1, y_2)$ and $\mathbf{Z}(z_1, z_2)$ be the end points of that segment. Since the remainder of line L is exterior to R, the end points $\mathbf{Y}$ and $\mathbf{Z}$ must be on the boundary of R. If the linear constraint defining the boundary through $\mathbf{Z}$ is $cx_1 + dx_2 \leq e$, then $cz_1 + dz_2 = e$, while $cw_1 + dw_2 = k < e$, since $\mathbf{W}$ is an interior point. The linear function $cx_1 + dx_2$ will assume values between k and e for points on L between $\mathbf{W}$ and $\mathbf{Z}$ by theorem 3–4. Similar strict inequalities will be satisfied by all the points between $\mathbf{W}$ and any other boundary line. Thus all points between an interior point and a boundary point are interior points. This means the open segment between $\mathbf{Y}$ and $\mathbf{Z}$ consists entirely of interior points and therefore L may intersect the boundary at only two points.

Let the given linear function be f (Figure 3–4). If $f(y_1, y_2) \leq f(z_1, z_2)$ then by theorem 3–4

$$f(y_1, y_2) \leq f(w_1, w_2) \leq f(z_1, z_2). \tag{1}$$

The boundaries of R are also closed segments by definition 3. The end points of boundary segments lie on exactly two boundary lines, so they are vertices. Thus if $\mathbf{Z}$ is not a vertex, it will lie between two vertices, say $\mathbf{S}$ and $\mathbf{T}$. Applying theorem 3–4 again, $f(z_1, z_2)$ is bounded on that boundary segment, say

$$f(t_1, t_2) \leq f(z_1, z_2) \leq f(s_1, s_2). \tag{2}$$

Combining inequalities (1) and (2) gives

$$f(w_1, w_2) \leq f(s_1, s_2). \tag{3}$$

---

[1] The existence is also guaranteed by a theorem from calculus. A linear function is continuous and any continuous function defined over a closed bounded region attains its extreme values at points of the region.

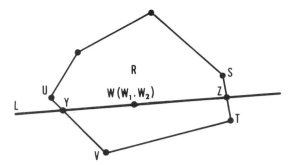

Figure 3-4

By the same argument $f(y_1, y_2)$ is bounded by the values of f at two vertices, say

$$f(v_1, v_2) \leq f(y_1, y_2) \leq f(u_1, u_2).\qquad(4)$$

Finally combine inequalities (1), (3), and (4), with the result

$$f(v_1, v_2) \leq f(w_1, w_2) \leq f(s_1, s_2).\qquad(5)$$

We have shown that f at any interior point **W** is bounded by the values of f on the boundary of R and those values in turn are bounded by values of f at vertices. Since there are only a finite number of vertices in a convex polygon, we may examine f at each vertex and choose one with as large a value of f as possible and one with as small a value of f as possible. These choices of vertices may not be unique, but they do represent maximum and minimum points of f over R, completing the proof. ■

In the case of a general polygonal convex region that may be unbounded, a linear function over that region may increase or decrease without bound. The intersection of a line through a point of R with R may be a ray or the entire line. Along a ray at least one coordinate must continue to increase or decrease and thus any linear f involving that coordinate may either increase or decrease without bound. To state a theorem in this case, it is necessary to hypothesize the existence of an optimum. We state the theorem without proof as the reasoning follows the same arguments given in theorem 3–5.

THEOREM 3-6. *If the maximum or the minimum value of a linear function defined over a polygonal convex region exists, then it is found on the boundary or at a vertex of the region.*

If a boundary were an entire line, then there would be no vertex on that boundary. However, for a feasible region in the first quadrant this could not happen, so every boundary line contains at least one vertex when coordinates are restricted to be nonnegative. As long as a vertex is available on the optimum boundary, the optimum value occurs at a vertex. Thus, the result is the same as in the case of a convex polygon, provided the optimum exists.

The proofs in this chapter have been carried out in two dimensions for simplicity and ease of understanding. The ideas generalize to n dimensions and hold for the general linear programming problem, where the variable subscripts run from 1 to n.

PROBLEMS

3–1. Find a set of parametric equations for the line segment from $P_1(2, -5)$ to $P_2(-3, 4)$. What are the parametric equations of the entire line through P_1 and P_2? For what values of parameter α do you get points on the line below P_1, and for what values of α do you get points on the line above P_2?

3–2. On line segment $P_1(-2, -4)$ to $P_2(7, 2)$, find the coordinates of the point P that is $2/3$ of the way from P_1 to P_2. Check your result by using the distance formula from analytic geometry.

3–3. Prove that an open half plane is a convex set by following the proof to Theorem 3–2.

3–4. For the linear function $f = 2x_1 + 3x_2$, defined over segment $P(-5, 3)$ to $Q(6, -1)$, find the maximum value of f and the minimum value of f. Find an arbitrary point interior to the segment PQ. Show that the value of f at this point is between the values of f at P and Q. Check your result by computing f as a function of α and testing it for your value of the parameter.

3–5. If $f = -3$ at $Y(-4, -1)$ and $f = 5$ at $Z(2, 3)$, find in terms of the parameter α the linear function $f(\alpha)$ defined over line segment YZ. Let (x_1, x_2) be an arbitrary point on segment YZ and find f as a function of x_1 and x_2. Note f is linear and may involve only one of the variables, either x_1 or x_2. By eliminating the parameter find the rectangular equation of the line through X and Y.

3–6. If $f(\alpha) = 3\alpha + 5$, $0 \le \alpha \le 1$, is an increasing function defined over the line segment from $Y(-5, 1)$ to $Z(2, -6)$, find the value of f at the end points of the segment. Find the coordinates of the point (x_1, x_2) for which $f = {}^{15}/_2$.

3–7. Write out the details to the proof of Theorem 3–6 by following the outline of the proof in Theorem 3–5.

3–8. Prove that a linear function defined over a convex polyhedron has both a maximum and a minimum value that occur at vertices of the polyhedron. Hint: Consider a plane section through any interior point and apply Theorem 3–5 to the resulting convex polygon.

3–9. Find the parametric equations of the line segment $P(1, 6)$ to $Q(5, -4)$. Find the coordinates of the point X that is on this segment $3/4$ of the way from P to Q. Given linear function $f = 3x_1 - 2x_2$, show that $f(P) < f(X) < f(Q)$.

3–10. In problem 3–9, find the given linear function in terms of parameter α. Does $f(\,^3/_4\,) = f(\,X\,)$? If f is defined on the entire line PQ, where are the points $f(\,\alpha\,)$ for $\alpha > 1$?

3–11. If a linear function has the values $f(\,0\,) = 3$ at point P and $f(\,1\,) = 7$ at point Q, find $f(\,\alpha\,)$. Find $f(\,^3/_4\,)$. Is a point on PQ corresponding to $\alpha = {}^3/_4$ closer to P or to Q?

3–12. Over segment $P(\,-3,\,5\,)$ to $Q(\,8,\,-1\,)$ function $f = 9\alpha - 2$ is defined for $0 \leq \alpha \leq 1$. Find the coordinates of point X on PQ for which $f(\,\alpha\,) = 4$.

3–13. In space R^3, find the parametric equations of the line segment $P(\,2,\,7,\,-3\,)$ to $Q(\,3,-8,\,7\,)$. Find the coordinates of the point X that is $^1/_5$ of the way from P to Q.

3–14. For the linear function $f = 2x_1 + 3x_2 + 4x_3$, find its maximum and minimum over the line segment in problem 3–13. Express f as an increasing function of α, $0 \leq \alpha \leq 1$, over this segment. In what direction over PQ does f increase?

3–15. If $f = 3x_1 + 4x_2$ is defined over the triangle with vertices $(\,1,\,2\,)$, $(\,5,\,6\,)$, and $(\,9,\,-1\,)$, then find the maximum value of f throughout this triangle. Find the minimum value of f over the same triangle. Evaluate f at a number of interior points such as $(5,5)$, $(6,5)$, $(7,2)$, and $(2,2)$ to compare with your maximum and minimum.

3–16. Suppose that $f = 3x_1 - 4x_2$. Answer the questions posed in problem 3–15. Do the extremes occur at different points than in problem 3–15?

CHAPTER 4

The Simplex Method

4.1 GAUSS-JORDAN ELIMINATION

The word *simplex* comes from the concept of the simplest possible convex figure that has exactly one more vertex than the dimension of the space. The zero-dimensional simplex is a single point, while a line segment, with two end points, is a simplex in one dimension. In two dimensions with three vertices, a simplex is a triangle, and in three dimensions with four vertices, a simplex is a pyramid or tetrahedron. Each of these figures is the smallest convex set containing its vertices, and is known as the convex hull of the vertices. By definition the *convex hull* of any set of points is the smallest convex set containing those points. Whenever the number of vertices or extreme points is finite, then the convex hull of those points is called a *convex polyhedron*. The simplices are then special cases of convex polyhedra having one more vertex than the space dimension, and with the additional property that their faces or boundaries are simplices of the next lower dimension. Linear programming has lifted the word simplex from geometry and applied it to the process of locating an optimal vertex among the vertices of a feasible region.

In developing the Simplex Method, George Dantzig made use of the classical Gauss-Jordan elimination process. Gauss-Jordan elimination for solving a system of linear equations was discussed in Chapter 1. The key idea is to take a multiple of one equation and add it to or subtract it from another equation to eliminate one of the unknowns from the second equation. The result of carrying out this elimination systematically is the row echelon form of the coefficients in the augmented matrix. The reduced system, whose solutions are easy to read off, is equivalent to the original system of linear equations. The following example will be carried out to illustrate the format and terminology used in linear programming.

Example 1.

$$2x_1 - 4x_2 + x_3 = 2$$
$$x_1 - 4x_2 - x_3 = 5$$

A notational convenience that is standard in linear programming is to place the coefficients of these equations into a rectangular block called a *tableau*. In Chapter 1 the rectangular array of numbers was called the *augmented matrix*.

x_1	x_2	x_3	
(2)	−4	1	2
1	−4	−1	5

To solve this system for unknowns x_1 and x_2, we will eliminate x_1 from the second equation and x_2 from the first equation. Then the first equation is immediately solvable for x_1 and the second equation is solvable for x_2. Each of these eliminations is called a *pivoting iteration* or, briefly, a *pivot* of the tableau. The initial elimination amounts to taking a multiple, in our case ½, of the first row and subtracting it from the second row. The operation revolves about the coefficient of x_1 in the first row. Thus this coefficient is called a *pivot* during the pivoting iteration. The pivots will be circled in our tableaux. The next tableau shows the first step in pivoting, which is to divide the row in which the pivot occurs (*pivotal row*) by the pivot.

x_1	x_2	x_3	
1	−2	½	1
1	−4	−1	5

The pivot position is now 1 in the above tableau. The final step in pivoting is to produce zeros in all remaining positions of the column in which the pivot occurs (*pivotal column*). In our case, subtract the first row from the second row.

$$
\begin{array}{ccc}
x_1 & x_2 & x_3
\end{array}
$$

x_1	x_2	x_3	
1	-2	$1/2$	1
0	(-2)	$-3/2$	4

The result of pivoting is a unit vector in the pivotal column.

The second pivoting iteration is to eliminate the coefficient of x_2 from the first row. Since x_2 is in the second column, the new pivot must be in the second column, but not in the first row, so choose the second row. If we were solving for x_3 instead of x_2, then the second pivot would be in the third column of the second row. Divide the second row by the pivot giving the following tableau.

x_1	x_2	x_3	
1	-2	$1/2$	1
0	1	$3/4$	-2

Then produce zeros in the remaining positions of the second column by taking twice the second row and adding it to the first row.

x_1	x_2	x_3	
1	0	2	-3
0	1	$3/4$	-2

This leaves a unit vector in the x_2 column.

The pivoting terminates when the number of unit column vectors is as large as possible. The number of unit vectors produced is equal to the rank of the matrix in the tableau. If this rank equals the number of rows m, then the unit vectors together compose the identity matrix, $\mathbf{I}'''$.

Each step in the pivoting process gives a system equivalent to the original system of equations, so each tableau possesses the same solution vectors. A solution for x_1 and x_2 in terms of x_3 may be read directly from the last tableau.

$$1x_1 + 0x_2 + 2x_3 = -3 \qquad\qquad x_1 = -3 - 2x_3$$

$$\text{or}$$

$$0x_1 + 1x_2 + {}^{3}/_{4}x_3 = -2 \qquad\qquad x_2 = -2 - {}^{3}/_{4}x_3. \quad \bullet$$

Of the many solutions, the most important for our purposes is the one in which $x_3 = 0$. This will be called a *basic solution* and is easily read off from the final tableau. The values of the *basic variables* are found in the right-hand column opposite the ones of the unit vectors. Thus $x_1 = -3$ and $x_2 = -2$ is a basic solution.

In the general case consider m equations in n unknowns where $m \leq n$. We will set $n - m$ of the variables equal to zero and try to solve for the remaining m variables.

DEFINITION 1.   For $m \leq n$ the $n - m$ variables that are chosen to be zero are called *nonbasic* and the remaining m variables are called *basic*. A solution for the basic variables is called a *basic solution*. A *nonbasic solution* is a solution to the problem that involves more nonzero variables than the original number of m equations.

A nonbasic solution vector in example 1 is $(-11, -5, 4)$ with three nonzero variables. Any solution with one variable equal to zero is a basic solution to this problem, so there are three basic solutions in which the basic variables are respectively (x_1, x_2), (x_1, x_3), and (x_2, x_3). The number of possible basic solutions may be found from the formula for the binomial coefficient

$$\binom{n}{m} = \frac{n!}{m!(n-m)!} \, .$$

A particular basic solution will exist and be unique, provided the matrix formed from the coefficients of those basic variables is nonsingular. This square m by m matrix of coefficients is nonsingular, provided its rank equals m.

4.2   THE EXTENDED TABLEAU

The tableaux in this section are called *extended tableaux* since they include an identity matrix. To illustrate the Simplex Method by use of the extended tableau consider the following linear program.

Example 2.   Find nonnegative numbers x_1 and x_2 subject to the constraints

$$x_1 + 3x_2 \leq 7$$
$$2x_1 + x_2 \leq 4,$$

such that the linear function $x_1 + 2x_2$ is a maximum. To convert the inequalities into equations, we introduce nonnegative *slack variables* x_3 and x_4. The slack variables take up the slack in the constraint. That is, they add into the left side whatever is necessary to bring it up to equality. The maximum will not be changed if the slack variables are given zero coefficients in the objective function. Write the objective as

$$x_1 + 2x_2 + 0x_3 + 0x_4 = M.$$

The problem may now be stated as follows. Find x_1, x_2, x_3, x_4 satisfying

$$x_1 \geq 0, x_2 \geq 0, x_3 \geq 0, x_4 \geq 0, \tag{1}$$

such that

$$x_1 + 3x_2 + x_3 + 0x_4 = 7$$
$$2x_1 + x_2 + 0x_3 + x_4 = 4,\tag{2}$$

and

$$M = x_1 + 2x_2 + 0x_3 + 0x_4 \tag{3}$$

is a maximum.

We are looking for an optimal feasible solution. Fortunately, we need only look among the basic solutions of system (2) to find our answer. The basic solutions of system (2) are the solutions for any two of the variables while the remaining two variables are set equal to zero. The two constraining equations in (2), plus the additional two equations setting the nonbasic variables equal to zero, constitute four faces or hyperplanes in the four-dimensional solution space of x_1, x_2, x_3, and x_4. The intersection of these four hyperplanes is a vertex or extreme point in the solution space with these hyperplanes as boundary faces. In general, the intersection of at least n hyperplanes in n-dimensional space is a vertex. The feasible region is that portion of the solution space satisfying the given constraints and the nonnegativity constraints (1). This feasible region is necessarily convex, since all constraints are linear and represent half spaces. The intersection of these half spaces is convex by theorem 3–1. Thus, the Extreme Point Theorem applies over the feasible region and the maximum (3) is found at a vertex determined by the four boundary hyperplanes. Since these vertices correspond to basic solutions, we will find our maximum among the basic solutions of (2) that satisfy (1). The result is stated in the following theorem.

THEOREM 4-1. *If an optimal feasible solution exists, then at least one basic optimal feasible solution exists as well.*

In the usual case, where the optimum is unique, theorem 4–1 guarantees that it will be basic. In case the optimum is not unique, we have what are called *alternate optima*. They need not be basic but at least one basic solution will have the same optimum value. As a matter of fact, unless the feasible region is unbounded, every nonbasic optimal feasible solution is a convex combination of two basic optimal feasible solutions.

Returning to our example, the initial extended tableau is

	x_1	x_2	x_3	x_4	
	1	③	1	0	7
	2	1	0	1	4
M	-1	-2	0	0	0

The coefficients of the objective function have been transferred to the left side of equation (3) and placed in a row at the bottom of our tableau. This row is referred to as the row of *cost coefficients* of the objective function. The box in the lower right corner will contain the objective value for the basic variables, provided those basic variables have zero coefficients in the bottom row. Choosing x_1 and x_2 to be nonbasic, $x_1 = x_2 = 0$, the initial basic feasible solution is $x_3 = 7$, $x_4 = 4$ with initial objective value $M = 0$. We may increase M by increasing any variable with a negative coefficient in the last row. The greatest increase in M per unit increase in x occurs by choosing the variable whose negative coefficient in the last row has the largest absolute value. This most negative entry is -2 and it determines the second column to be a pivotal column. We now try to find a pivotal row by asking the question, "How much may x_2 be increased?" Keep x_1 nonbasic, i.e., equal to zero, and consider the equations

$$x_3 = 7 - 3x_2$$

$$x_4 = 4 - 1x_2.$$

Since x_3 and x_4 must be nonnegative, it is clear that x_2 can be at most $7/3$. This is the smallest of the ratios $\theta_1 = 7/3, \theta_2 = 4/1$. The θ ratios are computed by dividing the entries in the right-hand column by the corresponding entries in the pivotal column. A pivotal row is then determined by choosing the smallest of the nonnegative ratios. In the interesting case of a tie for the smallest, either may be chosen. If we do not pick the smallest, then some variable will become negative and the resulting solution will be infeasible. For example, picking $\theta_2 = 4$ and increasing x_2 to 4 gives $x_3 = 7 - 12 = -5$. If an entry in the pivotal column is negative, the θ ratio will be negative, and the variable corresponding to that column may be increased without bound in that equation. For example, suppose θ_1 were negative, then $x_3 = 7 + 3x_2$ and x_3 remains feasible for any increase in x_2. If an entry in the pivotal column is zero, the θ ratio will be undefined and again no bound is placed on the corresponding x. Thus we consider θ only for the positive entries above some negative entry in the last row.

It is important to note that if all of the numbers above some negative number in the last row of a tableau are zero or negative, then the objective maximum is unbounded. Merely make the x of that column arbitrarily large. When this situation occurs, we will say that the problem has an *unbounded optimal solution*.

We have now chosen a pivotal column, the second, and a pivotal row, the first. Our pivot is entry 3 circled in the initial tableau. Pivoting on 3 performs the Gauss-Jordan elimination to give the following tableau.

	x_1	x_2	x_3	x_4	
	$1/3$	1	$1/3$	0	$7/3$
	$\boxed{5/3}$	0	$-1/3$	1	$5/3$
M	$-1/3$	0	$2/3$	0	$14/3$

This tableau shows that x_2 has come into the basis while x_3 went out. The new basic solution is $x_2 = 7/3$, $x_4 = 5/3$. We have increased x_2 from 0 to $7/3$ as predicted by the θ ratio, while M has increased from 0 to $14/3$. The presence of a negative in the last row indicates that we may pivot again, this time in the first column. Compute the θ ratios.

$$\theta_1 = 7/3 \cdot 3/1 = 7$$

$$\theta_2 = 5/3 \cdot 3/5 = 1$$

θ_2 is smaller and forces the pivotal row to be the second row. Carry out the pivoting on the entry $5/3$.

x_1	x_2	x_3	x_4	
0	1	$6/15$	$-1/5$	2
1	0	$-1/5$	$3/5$	1

	x_1	x_2	x_3	x_4	
M	0	0	$9/15$	$1/5$	5

No negative entries in the last row indicate that M may not be increased further. We have the final tableau and can read off the basic optimal feasible solution.

$$x_1 = 1$$

$$x_2 = 2$$

$$M = x_1 + 2x_2 = 1 + 4 = 5.$$

Theorem 4–1 assures us that no other feasible solution will have a larger value of M since no nonbasic feasible solution can have an objective value greater than that of all basic feasible solutions. ●

DEFINITION 2. An *extended tableau* is *feasible* if it contains an identity matrix, and the numbers above the objective row in the right-hand column are nonnegative.

A feasible tableau contains a basic feasible solution. Likewise a tableau is called *infeasible*, or *unbounded*, or *optimal* if it contains respectively a basic infeasible, or unbounded, or optimal solution.

The method of choosing a pivot may be summarized in the following algorithm known as the Simplex Algorithm.

THE SIMPLEX ALGORITHM

1. Start with a tableau that is feasible. If there are no negative numbers in the bottom row, then the tableau is already optimal. Otherwise proceed to step 2.

2. Choose a pivotal column by picking that column among the first $n-1$ columns whose entry in the bottom row is the most negative number. In case of ties, any one of the tying columns may be picked.

3. For the first $m-1$ rows compute the θ ratios for all positive entries in the pivotal column. (These are the ratios of the entry in the right-hand column to its corresponding entry in the pivotal column.) The θ ratios are all nonnegative.

4. Choose the pivotal row as one with the smallest θ ratio. In case of ties for the smallest, any one of the tying rows may be chosen.

5. Carry out a pivoting iteration using the pivot chosen by steps 2 and 4.

6. Repeat steps 2 through 5 until no new pivot can be found.

Normally the algorithm will terminate when either no pivotal column or no pivotal row can be found. That is, either there are no negative coefficients in the objective row, meaning a basic optimal feasible solution has been found, or above some negative coefficient there are no positive entries in the pivotal column, meaning an unbounded optimal solution exists.

It is worthwhile to compare our basic feasible solutions obtained by the simplex method with the graphical solution. The feasible region of the original problem is the intersection of four half planes as shown in Figure 4–1. The vertices of the convex polygon are (0, 0), (2, 0), (1, 2), and (0, $7/3$).

After slack variables are introduced the solution space is four dimensional. The equations in our tableaux are three-dimensional hyperplanes in four-space. The solution points are in the common two-dimensional space of these hyperplanes. If we project the solution set (x_1, x_2, x_3, x_4) onto the linear subspace (x_1, x_2, 0, 0) and associate this subspace with the two-space (x_1, x_2), then the set of feasible solutions can be considered to be the same as the convex set in Figure 4–1. Since the slack variables have zero coefficients in the objective function, we know that a solution will be projected onto (x_1, x_2, 0, 0) and is

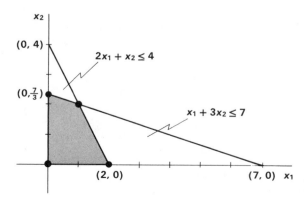

Figure 4–1

thus associated with a point (x_1, x_2). Making this association, we have the following solution points in our tableaux:

$$(x_1, x_2, x_3, x_4) \longrightarrow (x_1, x_2)$$

first B.F.S.	(0, 0, 7, 4)	$\longrightarrow$ (0, 0)
second B.F.S.	(0, 7/3, 0, 5/3)	$\longrightarrow$ (0, 7/3)
third B.F.S.	(1, 2, 0, 0)	$\longrightarrow$ (1, 2).

Note in Figure 4–1 that the process began at a vertex, the origin, and proceeded to adjacent vertices until the optimal vertex was reached. The simplex method is merely a way to move from one B.F.S. (basic feasible solution) to another with a value of **M** at least as great. Each move is equivalent to going from one vertex to an adjacent vertex of the feasible region. Since there are only a finite number of vertices, the optimal vertex must be reached in a finite number of steps, provided we improve the objective value at each step.

PROBLEMS

For the first three problems solve the given set of linear equations by pivoting down the main diagonal.

4–1. $3x - y + 6z = 1$

$x + 2y - 3z = 0$

$2x - 3y - z = -9$

4–2. $5x - y + 3z = 1$

$2x + 3y - z = -1$

$10x + 4y + 3z = -4$

4–3. $2x - 3y - z = 11$

$x + 4y - 2z = 3$

$3x + 4y + 5z = -5$

Solve the following linear programming problems by pivoting the extended tableau.

4–4. Find nonnegative numbers x_1 and x_2 satisfying the constraints

$x_1 + 2x_2 \le 10$

$3x_1 + 5x_2 \le 27,$

such that $11x_1 + 20x_2$ is a maximum. Check your answer graphically.

4–5. Find nonnegative numbers x_1 and x_2 satisfying the constraints

$$x_1 + 2x_2 \leq 8$$
$$2x_1 + 3x_2 \leq 13$$
$$x_1 + x_2 \leq 6,$$

such that $8x_1 + 9x_2$ is a maximum. Check your answer graphically.

4–6. Solve problem 2–14 to find the maximum profit of Seeall's Manufacturing Company.

4–7. Find nonnegative numbers x_1, x_2, and x_3 satisfying the constraints

$$3x_1 - 4x_2 + 8x_3 \leq 10$$
$$4x_1 - 2x_2 \qquad \leq 12$$
$$-x_1 + 3x_2 + 2x_3 \leq 7,$$

such that $3x_1 - x_2 - 2x_3$ is a maximum.

4–8. Caroline's Quality Candy Confectionary is famous for fudge, chocolate cremes, and pralines. Their candy making equipment is set up to make 100 lb. batches at a time. Currently there is a chocolate shortage and they can get only 120 lbs. of chocolate in the next shipment. On a week's run the confectionary's cooking and processing equipment is available for a total of 42 machine hours. During the same period the employees have a total of 56 man-hours available for packaging. A batch of fudge requires 20 lbs. of chocolate while a batch of cremes uses 25 lbs. of chocolate. The cooking and processing takes 120 minutes for fudge, 150 minutes for chocolate cremes, and 200 minutes for pralines. The packaging times measured in minutes per pound box are 1, 2, and 3 respectively for fudge, cremes, and pralines. How many batches of each type of candy should the confectionary make, assuming that the profit per pound box is $1.50 on fudge, $1.40 on chocolate cremes, and $1.45 on pralines. Also find the maximum profit for the week.

4.3 THE CONDENSED TABLEAU

Example 3. Consider again example 2 in section 4.2. Find $x_1 \geq 0$, $x_2 \geq 0$ such that

$$x_1 + 3x_2 \leq 7$$
$$2x_1 + x_2 \leq 4$$

and $x_1 + 2x_2 = M$ is a maximum. Note that the columns of unit vectors in the extended tableau offer no significant information other than to designate the basic variables. It will be convenient to omit these columns and place the subscripts of the basic variables in a column to the left of our tableau. These subscripts will indicate the row of that basic variable and its corresponding basic value will be found in the right-hand column. Finally the columns will be labeled at the top of our tableau with the subscripts of the corresponding nonbasic variables. The initial *condensed tableau* that appears below contains all of the same information as the corresponding extended tableau while omitting its unit column vectors.

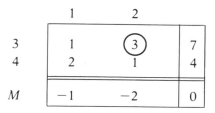

Notice how easily this tableau may be set up from the original constraints and the basic solution read off, $x_3 = 7$, $x_4 = 4$, and $M = 0$. As before, the pivot is determined by the θ ratios of the second column, $\theta_1 = {}^7/_3, \theta_2 = {}^4/_1$, to be the number 3 circled above. Now look at the second tableau of the previous section and condense it into the following.

	1	3	
2	$^1/_3$	$^1/_3$	$^7/_3$
4	$^5/_3$	$-^1/_3$	$^5/_3$
M	$-^1/_3$	$^2/_3$	$^{14}/_3$

This tableau contains all of the necessary information, and the second B.F.S. is read off immediately as $x_2 = {}^7/_3$, $x_4 = {}^5/_3$, $M = {}^{14}/_3$.

The secret is to determine this second tableau directly from the first condensed tableau without resorting to the extended tableau. A close comparison of the two condensed tableaux reveals the following five step algorithm.

THE CONDENSED TABLEAU PIVOTING ALGORITHM

1. Replace the pivot with its reciprocal.
2. Divide the remaining entries of the pivotal row by the pivot.
3. Divide the remaining entries of the pivotal column by the negative of the pivot.
4. For the entries in the remaining positions, subtract from the corresponding position in the old tableau the product of the old value both in that

row and in the pivotal column times the new entry both in that column and in the pivotal row.

5. Interchange the integer above the pivotal column with the integer to the left of the pivotal row.

Try these five steps out on the second tableau, pivoting on the number $5/3$, to check the final tableau.

	1	3	
2	$1/3$	$1/3$	$7/3$
4	$5/3$	$-1/3$	$5/3$
M	$-1/3$	$2/3$	$14/3$

	4	3	
2	$-1/5$	$6/15$	2
1	$3/5$	$-1/5$	1
M	$1/5$	$9/15$	5

To illustrate step 4, suppose the matrix entries are labeled Y_{ij}, i running from 1 to $m = 3$, j running from 1 to $n = 3$. Let the pivotal row be p and the pivotal column be q so that the pivot is Y_{pq}. Then any new entry $\overline{Y}_{ij}$ that is neither in the pivotal row nor in the pivotal column is found according to step 4 by the formula

$$\overline{Y}_{ij} = Y_{ij} - Y_{iq}\frac{Y_{pj}}{Y_{pq}}$$

To compute the final value of M we have

$$\overline{Y}_{33} = Y_{33} - Y_{31}\frac{Y_{23}}{Y_{21}} = 14/3 - (-1/3)\frac{5/3}{5/3} = 5. \quad \bullet$$

The whole process is readily programmed for a digital computer and such a program is given in the next chapter.

To prove that the steps of our algorithm hold, condense the tableaux of a pivot of the general extended tableau and then compare these condensed tableaux. The general extended tableau is indicated schematically below where variable s is chosen to come into the basis replacing basic variable u. The pivot is Y_{pq}.

$$
\begin{array}{ccc}
\text{pivotal} & \text{some} & \\
\text{column} & \text{other} & \\
s & \text{column} & u
\end{array}
$$

	pivotal column s	some other column	u
pivotal row	$\ldots \; Y_{pq} \; \ldots$	$Y_{pj} \; \ldots \; 1 \; \ldots$	Y_{pn}
some other row	$\ldots \; Y_{iq} \; \ldots$	$Y_{ij} \; \ldots \; 0 \; \ldots$	Y_{in}

The extended tableau pivoting process produces the following tableau where s is now basic.

s		u	
$\ldots \; 1 \; \ldots$	Y_{pj}/Y_{pq}	$\ldots \; 1/Y_{pq} \; \ldots$	Y_{pn}/Y_{pq}
$\ldots \; 0 \; \ldots$	$Y_{ij} - Y_{iq}(\; Y_{pj}/Y_{pq} \;) \; \ldots$	$-Y_{iq}/Y_{pq} \; \ldots$	$Y_{in} - Y_{iq}(\; Y_{pn}/Y_{pq} \;)$

Condense the first tableau by dropping all columns of unit vectors. Note u appears to the left opposite its value Y_{pn}.

		pivotal column s	some other column	
pivotal row	u	$\ldots \; Y_{pq} \; \ldots$	$Y_{pj} \; \ldots$	Y_{pn}
some other row		$\ldots \; Y_{iq} \; \ldots$	$Y_{ij} \; \ldots$	Y_{in}

In condensing the second extended tableau, replace the s column with the u column and drop all other columns of unit vectors.

$$
\begin{array}{|c c c|c|}
\hline
 & \vdots & \vdots & \vdots \\
s & \ldots \; 1/Y_{pq} \; \ldots & Y_{pj}/Y_{pq} \qquad \ldots & Y_{pn}/Y_{pq} \\
 & \vdots & \vdots & \vdots \\
 & \ldots -Y_{iq}/Y_{pq} \; \ldots \; Y_{ij}-Y_{iq}(\,Y_{pj}/Y_{pq}\,) \ldots & & Y_{in}-Y_{iq}(\,Y_{pn}/Y_{pq}\,) \\
 & \vdots & \vdots & \vdots \\
\hline
\end{array}
$$

The five step algorithm is clearly evident by comparing these condensed tableaux.

The Simplex Algorithm for choosing a pivot applies as well to the condensed tableau as it did to the extended tableau. The only difference is that the pivoting is now carried out by the Condensed Tableau Pivoting Algorithm.

PROBLEMS

Solve the following linear programs by pivoting the condensed tableau.

4–9. Find $x_1 \geq 0$ and $x_2 \geq 0$ such that

$$2x_1 + 3x_2 \leq 30$$
$$x_1 + x_2 \leq 11$$
$$4x_1 + 3x_2 \leq 40,$$

and $5x_1 + 6x_2$ is a maximum.

4–10. Find $x_1 \geq 0$ and $x_2 \geq 0$ such that

$$2x_1 + 3x_2 \leq 30$$
$$x_1 + x_2 \leq 11$$
$$4x_1 + 3x_2 \leq 40,$$

and $7x_1 + 6x_2$ is a maximum. Check your answer graphically.

4–11. Find $x_1 \geq 0$, $x_2 \geq 0$, and $x_3 \geq 0$ such that

$$x_1 + x_2 + 2x_3 \leq 200$$
$$2x_1 + x_2 \leq 100$$
$$10x_1 + 8x_2 + 5x_3 \leq 2000,$$

and $2x_1 + 4x_2 + x_3$ is a maximum.

4-12. Find $x_1 \geq 0$ and $x_2 \geq 0$ such that

$$5x_1 + 2x_2 \leq 45$$
$$-4x_1 + 5x_2 \leq 30,$$

and $-2x_1 + 5x_2$ is a maximum. Find the minimum of the same objective function. Hint: Let the minimum be the negative of a maximum, $m = -M$, and then maximize the expression M. Check your answers graphically.

4-13. Find $x_1 \geq 0$ and $x_2 \geq 0$ such that

$$x_1 - x_2 \geq -3$$
$$3x_1 + 4x_2 \leq 40$$
$$x_1 \leq 9$$
$$7x_1 + 2x_2 \leq 64,$$

and $4x_1 + x_2$ is a maximum. Hint: How can you reverse the sense of the first inequality?

4-14. With the same constraints as problem 4-13, maximize $3x_1 + x_2$.

4-15. With the same constraints as problem 4-13, maximize $x_1 - 2x_2$.

4-16. With the same constraints as problem 4-13, minimize $x_1 - 2x_2$. Note the hint in problem 4-12.

4-17. A small boat manufacturer builds three types of fiberglass fishing boats, a pram, a run-a-bout, and a trihull whaler. He sells the pram at a profit of $175, the run-a-bout at a profit of $190, and the whaler at a profit of $200. The factory is divided into two sections. Section A does the molding and construction work while section B does the painting, finishing, and equipping. The pram takes 1 hour in section A and 2 hours in section B. The run-a-bout takes 2 hours in section A and 5 hours in section B. The whaler takes 3 hours in section A and 4 hours in section B. Shop A has a total of 6240 hours available and shop B has a total of 10,800 hours available for the year. The manufacturer has ordered a year's supply of fiberglass that is sufficient to build at most 3000 boats figuring the average amount of fiberglass used per boat. What is his year's production schedule and profit in order to earn the maximum profit?

4-18. Suppose in problem 4-17 that the run-a-bout sold for a profit of $200 and the whaler sold for a profit of $190, while the profit on the pram remained at $175. Then what should the production schedule and maximum profit be?

4.4   ARTIFICIAL VARIABLES

So far in Chapter 4 we have considered only type I inequalities ($\leq$). A linear program may have constraints utilizing type II inequalities ($\geq$) or even equalities. In the latter two cases artificial variables are introduced to find an initial basic feasible solution. Let us consider the following problem which illustrates all three types of constraints.

Example 4.   The Navy is experimenting with three types of bombs, A, B, and C, in which three kinds of explosives will be used, D, E, and F. The Captain wishes to use exactly 2000 pounds of explosive D, at least 1000 pounds of explosive E, and at most 3000 pounds of explosive F. Bomb A requires 3, 2, 1 pounds of D, E, and F respectively. Bomb B requires 1, 5, 2 pounds of D, E, and F respectively. Bomb C requires 6, 1, 4 pounds of D, E, and F respectively. Now bomb A will give the equivalent of a 1 ton explosion, bomb B will give a 4 ton explosion, and bomb C will give a 3 ton explosion. Under what production schedule can the Navy make the biggest bang?

SOLUTION.   First, choose the unknowns x_1, x_2, and x_3 for the number of bombs to be produced of types A, B, and C respectively. The constraints on the amount of each kind of explosive available give the following system

$$3x_1 + 1x_2 + 6x_3 = 2000$$
$$2x_1 + 5x_2 + 1x_3 \geq 1000 \tag{1}$$
$$1x_1 + 2x_2 + 4x_3 \leq 3000.$$

The objective is to create the largest explosive force, so

$$1x_1 + 4x_2 + 3x_3 = M$$

should be maximized subject to constraints (1) and the nonnegativity requirement $x_1 \geq 0$, $x_2 \geq 0$, $x_3 \geq 0$.

Let us set up the extended tableau and then condense it. We introduce slack variables x_4 and x_7 to convert the inequalities into equalities. System (1) now appears as

$$3x_1 + 1x_2 + 6x_3 \qquad = 2000$$
$$2x_1 + 5x_2 + 1x_3 - 1x_4 = 1000 \tag{2}$$
$$1x_1 + 2x_2 + 4x_3 + 1x_7 = 3000.$$

The requirement that the slack variables also be nonnegative forces the coefficient of x_4 to be negative. In this case the slack variable represents an excess that must be subtracted from the left side of the inequality to reduce it to equality. Unfortunately we have only one column unit vector and no basic feasible solution. In order to gain two more column unit vectors, we introduce two *artificial variables* $x_{-5} \geq 0$ and $x_{-6} \geq 0$ into the first two equations of (2). The subscripts are made negative so that a computer may easily recognize artificial variables. To keep from losing track of the original problem while solving the *artificial problem*, we charge an arbitrarily high cost for the artificial variables in the objective function. Let the objective function of the artificial problem be

$$\overline{M} = M - N(x_{-5} + x_{-6})$$

where N is a large positive number. Since an arbitrarily large amount is subtracted from M, $\overline{M}$ cannot possibly reach a maximum unless $x_{-5} = x_{-6} = 0$. Due to this high cost N, pivoting will drive the artificial variables out of the basis forcing them to zero and leaving $\overline{M} = M$. The artificial problem can be stated as follows: Find x_1, x_2, x_3, x_4, x_{-5}, x_{-6}, x_7 all ≥ 0 such that

$$3x_1 + 1x_2 + 6x_3 + 1x_{-5} \qquad\qquad = 2000$$
$$2x_1 + 5x_2 + 1x_3 - 1x_4 \quad + 1x_{-6} = 1000$$
$$1x_1 + 2x_2 + 4x_3 \qquad\qquad + 1x_7 = 3000,$$

and $1x_1 + 4x_2 + 3x_3 - N(x_{-5} + x_{-6}) = \overline{M}$ is a maximum.

To separate the arbitrarily large coefficients from the relatively small coefficients, create a double row for the objective function where the bottom row of the tableau represents the multiples of N. Here is the initial extended tableau.

	x_1	x_2	x_3	x_4	x_{-5}	x_{-6}	x_7	
	3	1	6	0	1	0	0	2000
	2	5	1	−1	0	1	0	1000
	1	2	4	0	0	0	1	3000
$\overline{M}$	−1	−4	−3	0	0	0	0	0
	0	0	0	0	1	1	0	0

The 1s must now be removed from the bottom row so that x_{-5}, x_{-6}, x_7 and M will form a basic solution. This is easily done by multiplying each of the first two equations by N and subtracting from the objective equation.

	x_1	x_2	x_3	x_4	x_{-5}	x_{-6}	x_7	
	3	1	6	0	1	0	0	2000
	2	5	1	−1	0	1	0	1000
	1	2	4	0	0	0	1	3000
$\overline{M}$	−1	−4	−3	0	0	0	0	0
	−5	−6	−7	1	0	0	0	−3000

The first basic feasible solution is read off as $x_{-5} = 2000$, $x_{-6} = 1000$, and $x_7 = 3000$ while the remaining variables are zero. The starting value of $\overline{M}$ is a large negative number, $-3000N$. Dropping the column unit vectors we have the following initial condensed tableau.

	1	2	3	4	
−5	3	1	⑥	0	2000
−6	2	5	1	−1	1000
7	1	2	4	0	3000
$\overline{M}$	−1	−4	−3	0	0
	−5	−6	−7	1	−3000

The pivoting is carried out by choosing a pivotal column according to negatives in the last row as long as nonzero artificial variables remain in the basis. As soon as an artificial variable is removed from the basis, that column may be dropped from the tableau. Since a maximum for $\overline{M}$, if it exists, requires that all artificial variables go to zero, we will never pivot again on a column headed by such a variable. Checking the θ ratios for the third column, above entry −7, we find entry 6 for the first pivot. The resulting tableaux listed below were found on a computer.

	1	2	−5	4	
3	.5	.167	.	0	333.33
−6	1.5	④.83	.	−1	666.67
7	−1	1.33	.	0	1666.67
$\overline{M}$	.5	−3.5	.	0	1000.
	−1.5	−4.83	.	1	−666.67

The third column is dropped, and the second pivot is found to be entry 4.83 to obtain the next tableau. The artificial variables are now out of the basis and may be neglected. Thus, the second column may be dropped. Notice that the remaining entries in the last row are all zeros and in particular $\overline{M} = M$ is no longer affected by the last row. Dropping the last row completely, we next enter phase two of a two phase problem.

	1	−6	4	
3	.448	.	.034	310.35
2	.310	.	−.207	137.93
7	−1.414	.	(.276)	1482.76
$\overline{M}$	1.586	.	−.724	1482.76
	0	.	0	0

Eliminating the artificial variables is often called *phase one* and then the second phase continues the pivoting according to negative entries in the first of the two objective rows. Our third pivot is thus found above −.724 to be .276.

	1	7	
3	(.625)	−.125	125.
2	−.750	.750	1250.
4	−5.125	3.625	5375.
M	-2.125	2.625	5375.

The fourth pivot must be .625 because that is the only positive entry above the negative entry −2.125. This final pivot gives the optimal tableau listed below.

	3	7	
1	1.6	−.2	200
2	1.2	.6	1400
4	8.2	2.6	6400
M	3.4	2.2	5800

No further pivoting is possible because the last row is all positive. The final solution is read off to be $x_1 = 200$, $x_2 = 1400$, $x_3 = 0$, $x_4 = 6400$, $x_7 = 0$, and $M = 5800$. The Navy should produce 200 bombs of type A, 1400 bombs of type B, and none of the type C bombs for a total explosive power of 5800 tons. Since slack variable $x_7 = 0$, all 3000 pounds of explosive F will be used. Slack variable $x_4 = 6400$ means that an additional 6400 pounds of explosive E is utilized above the required 1000 pounds. These results can all be checked in the original constraints. ●

4.5 MINIMIZATION

The previous examples in this chapter have called for maximizing the objective function. It is convenient to convert every linear programming problem into finding a maximum so that one computer routine will handle both cases. This can easily be done. If the problem calls for minimizing the objective function, then maximize the negative of the objective function. This essentially reflects the problem with respect to the origin and the desired minimum is the negative of the maximum found in the final tableau.

Example 5. Consider example 4 in section 4.4 with a new objective. Under the same constraints on the amounts of explosive to be used, find the production schedule that will give the minimum cost of explosive ingredients. Suppose explosive D costs \$100 per lb., explosive E costs \$500 per lb., and explosive F costs \$300 per lb. Find the minimum cost.

SOLUTION.   Select two slack variables, x_4, x_7, and two artificial variables, x_{-5}, x_{-6}. The slack variables have zero cost and the artificial variables have a very large cost, N. The three types of bombs, A, B, and C, will cost \$1600, \$3200, and \$2300 respectively. The objective cost equation is thus

$$m = 1600x_1 + 3200x_2 + 2300x_3. \tag{1}$$

First set $m = -M$ where M is to be maximized. Then set up the artificial problem to maximize $\overline{M}$ where

$$\overline{M} = M - N(x_{-5} + x_{-6}). \tag{2}$$

By combining (1) and (2) the objective is

$$\overline{M} + 1600x_1 + 3200x_2 + 2300x_3 + N(x_{-5} + x_{-6}). = 0. \tag{3}$$

The initial extended tableau is given here.

	x_1	x_2	x_3	x_4	x_{-5}	x_{-6}	x_7	
	3	1	6	0	1	0	0	2000
	2	5	1	−1	0	1	0	1000
	1	2	4	0	0	0	1	3000
$\overline{M}$	1600	3200	2300	0	0	0	0	0
	0	0	0	0	1	1	0	0

Once again we must eliminate the 1s in the last row by subtracting N times the first two rows, giving

	x_1	x_2	x_3	x_4	x_{-5}	x_{-6}	x_7	
	3	1	6	0	1	0	0	2000
	2	5	1	−1	0	1	0	1000
	1	2	4	0	0	0	1	3000
$\overline{M}$	1600	3200	2300	0	0	0	0	0
	−5	−6	−7	1	0	0	0	−3000

Eliminating unit vectors, the initial condensed tableau is

	1	2	3	4	
−5	3	1	⑥	0	2000
−6	2	5	1	−1	1000
7	1	2	4	0	3000
$\overline{M}$	1600	3200	2300	0	0
	−5	−6	−7	1	−3000

With a little skill and practice one can set up the condensed tableau directly without resorting to the extended tableau.

The first two pivots occur at the same entries as in the previous problem. The pivots are circled in each tableau. Every time an artificial variable is removed from the basis, that column is dropped. The problem is now solved on a computer that lists the following sequence of three tableaux.

	1	2	−5	4	
3	.5	.167	.	0	333.33
−6	1.5	④.833	.	−1	666.67
7	−1.	1.333	.	0	1666.67
$\overline{M}$	450.	2817.	.	0	−766667.
	−1.5	−4.833	.	1	−666.67

	1	−6	4	
3	.448	.	.0345	310.35
2	⓪.310	.	−.207	137.93
7	−1.414	.	.276	1482.76
$\overline{M}$	−424.1	.	582.8	−1155170.
	0	.	0	0

	2	4	
3	−1.444	.333	111.11
1	3.222	−.667	444.44
7	4.556	−.667	2111.11
M	1366.7	300.	−966667.

Our problem is completed on three pivots and the answer read off from the final tableau. As frequently happens, the answers are not integral. This difficulty may be resolved by Integer Programming (discussed in Chapter 11.) At this time, however, we accept the closest integer as an approximate answer. The Navy should make 444 of the type A bombs, none of the type B bombs, and 111 of the type C bombs at a minimum cost of \$966,667. Since slack variable $x_4 = 0$, only 1000 pounds of explosive E are used. The slack $x_7 = 2111$ indicates that 2111 pounds of the 3000 pounds of explosive F remain unused. If these results are checked in the original constraints, the discrepanices due to rounding off will appear. ●

The bottom row of the double row objective function in the condensed tableau is the negative of the sum of the artificial variables. The artificial problem or phase one of the original problem can be considered a separate linear program. In the phase one program, the objective is to minimize the sum of the artificial variables or to maximize the negative of this sum. If the phase one program terminates with its objective equal to zero, then each of the artificial variables must be zero since they are nonnegative. When all artificial variables are zero, the phase two program begins with a B.F.S. to the original problem.

4.6   UNBOUNDEDNESS AND INCONSISTENCY

A linear programming problem may not have a solution. This situation should be detectable from the final tableau. The pivoting procedure continues to produce basic feasible solutions until no new pivot can be found. The process might stop because no pivotal row can be found to go with a possible pivotal column. This case occurs when there are no positive entries above some negative entry in the last row. The variable corresponding to this column could then be brought into the basis at an arbitrarily large value causing the objective function to be unbounded. The situation is easily noted in the final tableau as illustrated in example 6.

Example 6.   Find nonnegative numbers x_1 and x_2 such that

$$x_1 - 2x_2 \leq 2$$
$$3x_1 - 2x_2 \leq 18,$$

and $2x_1 - x_2$ is a maximum. The initial tableau and two pivots are given below

	1	2	
3	①	−2	2
4	3	−2	18
M	−2	1	0

Pivoting at $p, q = 1, 1$:

	3	2	
1	1	−2	2
4	−3	④	12
M	2	−3	4

Pivoting at $p, q = 2, 2$:

	3	4	
1	−1/2	1/2	8
2	−3/4	1/4	3
M	−1/4	3/4	13

We have arrived at the final tableau since no pivotal row can be found. However a negative in the objective row indicates that a maximum has not been reached. If we were to bring x_3 back into the basis, the equations which should limit the value of x_3 are

$$x_1 = 8 + \tfrac{1}{2}x_3 \qquad\qquad\qquad\qquad (1)$$
$$x_2 = 3 + \tfrac{3}{4}x_3.$$

The value of x_3 may be arbitrarily large in equations (1) without destroying the feasibility of x_1 and x_2. Consequently x_1 and x_2 may be arbitrarily large subject only to

$$x_1 = \tfrac{2}{3}x_2 + 6.$$

The result is clearly illustrated in Figure 4–2.

The feasible region is the unbounded shaded area in Figure 4–2. The three basic feasible solutions found in the above tableaux are respectively the

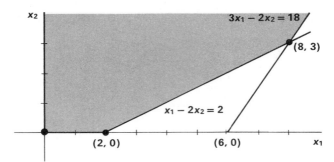

Figure 4–2

vertices (0, 0), (2, 0), and (8, 3). By continuing from vertex (8, 3) upward along the boundary $3x_1 - 2x_2 = 18$, the objective function may be increased without bound. If $x_1 = 100$ then $x_2 = 141$, and the objective is $M = 2(100) - 141 = 59$. The final tableau must be viewed carefully to see if a maximum has truly been reached. ●

A solution may also not exist if the original constraints are inconsistent. This happens when a nonzero artificial variable cannot be removed from the basis. A nonzero basic artificial variable in the final tableau signifies that we cannot find a basic feasible solution to the original problem. The artificial variables were introduced for the sole purpose of finding an initial basic feasible solution to the original problem and they must go to zero to get such a solution. Thus, if a nonzero artificial variable remains in the final tableau, there are no feasible points and the original constraints must be inconsistent. Consider example 7.

Example 7. Find $x_1 \geq 0$ and $x_2 \geq 0$ such that

$$2x_1 + 5x_2 \leq 10$$
$$-4x_1 + 7x_2 \geq 21,$$

and $x_1 + 2x_2$ is a maximum. To set up the initial tableau let x_3 and x_4 be the slack variables, and let x_{-5} be the artificial variable for the type II constraint. Let $M = x_1 + 2x_2$ and let the maximum of the artificial problem be $M = \overline{M} - Nx_{-5}$. Then the new problem appears as follows. Find $x_1, x_2, x_3, x_4, x_{-5}$ all ≥ 0 such that

$$2x_1 + 5x_2 + x_3 = 10$$
$$-4x_1 + 7x_2 - x_4 + x_{-5} = 21,$$

and $\overline{M} = x_1 + 2x_2 - Nx_{-5}$ is a maximum.

The initial condensed tableau with a double row for the objective function is given next. Notice that in order to get x_{-5} into the basis, the second row has been subtracted from the last row.

	1	2	4	
3	2	⑤	0	10
−5	−4	7	−1	21
$\overline{M}$	−1	−2	0	0
	4	−7	1	−21

Pivoting on 5 at matrix position (1, 2) produces the next and final tableau.

	1	3	4	
2	.4	.2	0	2
−5	−6.8	−1.4	−1	7
$\overline{M}$	−.2	.4	0	4
	6.8	1.4	1	−7

Since no negative numbers are in the last row, there is no possible pivotal column. The final basic solution is $x_2 = 2$, $x_{-5} = 7$. $\overline{M} = 4 − 7N$ where N is arbitrarily large. Thus $\overline{M}$ is an arbitrary negative number and never reaches a maximum. The artificial problem never reduces to the original problem. There-fore, the original problem has no solution and its constraints must be inconsist-ent. The inconsistency shows up on the following graph, Figure 4–3. It shows that there are no feasible points. ●

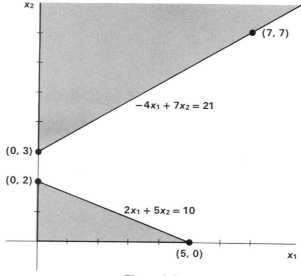

Figure 4–3

4.7 DEGENERACY

Another difficulty in the simplex method arises when one of the basic variables is zero in addition to the nonbasic variables that are zero by definition. If a basic variable becomes zero in a given tableau, then one of the θ ratios used for finding the next pivotal row may be zero. Since zero is the minimum possible θ ratio, the next pivotal row would be that one with the zero basic variable. The value of the minimum θ ratio is also the value of the new variable introduced into the basis. Using step 2 of our algorithm for pivoting the condensed tableau, this new variable comes into the basis at the old value divided by the pivot. Thus the new variable introduced will remain at value zero. By checking step 4 of the algorithm

$$\overline{Y}_{in} = Y_{in} - Y_{tq} Y_{pn} / Y_{pq},$$

we see that when $Y_{pn} = 0$, Y_{in} remains unchanged for all i. Thus all basic variables in the new tableau have the same value that they held in the old tableau. The only difference in the final column of the two tableaux is that the basic variable with value zero has been renamed. In particular the objective value has not changed. We will call such a basic feasible solution degenerate.

DEFINITION 3.   A basic feasible solution that has one or more basic variables equal to zero is called *degenerate*.

THEOREM 4-2.   *Degeneracy will occur whenever there is a tie for least among the θ ratios used to determine a pivotal row.*

PROOF.   Consider the condensed tableau

$$
\begin{array}{c|c}
\begin{matrix} \vdots & & \vdots \\ \therefore \ldots & Y_{pq} & \ldots \\ & \vdots & \\ & \vdots & \\ \ldots & Y_{rq} & \ldots \\ & \vdots & \end{matrix} &
\begin{matrix} \vdots \\ Y_{pn} \\ \vdots \\ \vdots \\ Y_{rn} \\ \vdots \end{matrix}
\end{array}
$$

and suppose there is a tie between the pth and r^{th} rows for the least θ ratio. Then,

$$Y_{pn} / Y_{pq} = Y_{rn} / Y_{rq} \tag{1}$$

Choose Y_{pq} as pivot and carry out the pivoting algorithm.

$$
\begin{array}{c|c}
\vdots & \vdots \\
\cdots \quad 1/Y_{pq} \quad \cdots & Y_{pn}/Y_{pq} \\
\vdots & \vdots \\
\cdots \quad -Y_{rq}/Y_{pq} \quad \cdots & Y_{rn} - Y_{rq}(Y_{pn}/Y_{pq}) \\
\vdots & \vdots
\end{array}
$$

From equation (1)

$$Y_{rn} - Y_{rq} Y_{pn} / Y_{pq} = 0.$$

Thus, a basic variable is zero and the corresponding basic feasible solution is degenerate. ■

In most practical cases degeneracy does not cause any difficulties. Pivoting continues until an optimal tableau is reached. Because the objective value remains constant through a number of pivots, different sets of basic variables will have the same objective value. The solution can still be unique, because the values of the variables don't change while the objective function remains constant. There is a theoretical difficulty. If the objective value cannot be improved at each pivot, then we cannot be sure that a final answer will ever be reached. On the other hand, as long as the objective is improved with each pivot, we must reach the optimum in a finite number of steps since only a finite number of basic solutions exist. In the degenerate case, it is possible to get into a cycle while the objective function is constant. This means that after a number of pivots you return to a previously encountered basic solution. Of course this cycle would be disastrous on a computer. The machine would be in an infinite loop with no way to get out.

Problems that cycle have occurred in practice as reported by P. Wolfe (1963) and by T. C. T. Kotiah and D. I. Steinberg (1978). Their problems cycled in periods of 25 and 18 iterations respectively. Fortunately, such cases are rare.

The following example is from E. M. L. Beale:[1]

---

[1] E. M. L. Beale, "Cycling in the Dual Simplex Algorithm," *Naval Research Logistics Quarterly*, 2, 1955: 269–275.

Example 8.

	1	2	3	4	
5	$1/4$	-8	-1	9	0
6	$1/2$	-12	$-1/2$	3	0
7	0	0	1	0	1
	$-3/4$	20	$-1/2$	6	0

If all ties in this example are broken by choosing the basic variable of lowest subscript to go out of the basis, then there is a cycle of period six. The basic variables in the cycle are: (5, 6, 7), (1, 6, 7), (1, 2, 7), (3, 2, 7), (3, 4, 7), (5, 4, 7), (5, 6, 7). To get out of the cycle it is necessary to break the ties in some other order. One can write a computer program that will break all ties in such a way to make cycling impossible. Several of these techniques are discussed in the next section. ●

4.8 BREAKING A CYCLE

Cycling can only occur if the problem is degenerate. In order for a B.F.S. to be repeated after a number of pivoting iterations, the objective value must remain the same. This can only happen when the right-hand constant in the pivotal row is zero, that is, the problem is degenerate. If degeneracy can be avoided, then the objective value must increase with each pivot and the corresponding basic feasible solutions are distinct.

Degeneracy may be avoided in several ways. One is called *perturbation*. The idea is to perturb the right-hand column by adding a set of small positive numbers to the Y_{in}, $i = 1, 2, \ldots, m-1$, in order that all ties among θ-ratios are broken in a unique manner. The simplest case is to choose the epsilon numbers as ε, ε^2, $\varepsilon^3, \ldots$, ε^{m-1}. The perturbed column is then $Y_{1,n} + \varepsilon$, $Y_{2,n} + \varepsilon^2, \ldots$, $Y_{m-1,n} + \varepsilon^{m-1}$. By reversing the proof to theorem 4–2, it may be shown that the only way for a nondegenerate tableau to become degenerate is for tying θ-ratios to exist. If the basic variable in the rth row becomes zero after pivoting, then

$$Y_{rn} - Y_{rq} Y_{pn} / Y_{pq} = 0$$

or

$$Y_{rn} / Y_{rq} = Y_{pn} / Y_{pq}.$$

Thus,

$$\theta_r = \theta_p$$

and the previous tableau had a tie for the pivotal row. Perturbation both eliminates degeneracy in the initial tableau and eliminates any possible ties among θ-ratios in subsequent tableaux, so that degeneracy and cycling are impossible. When the final tableau of a perturbed problem is reached, the solution to the original problem is found by setting $\varepsilon = 0$.

Another way of breaking ties between possible pivotal rows and avoiding cycling is called *lexicographic ordering*. The term *lexicographic* comes from the ordering of words in a dictionary by the first letter that is distinct in each word. Two sequences of numbers are ordered lexicographically by the first pair of numbers from each sequence that are distinct. Mathematically the lexicographic method is equivalent to the perturbation method. Lexicographic ordering avoids the question of how small ε must be chosen to ensure that the solution to the unperturbed problem is not changed.

A more recent rule for breaking cycles was developed by R.G. Bland[2] in 1977. The rule by Bland, called the *Smallest Subscript Rule*, is especially easy to implement on the computer. It has two steps, one for choosing the entering variable, and one for choosing the departing variable in each iteration. The Smallest Subscript Rule follows:

1. Among all of the columns with negative cost coefficients in the objective row, pick for the pivotal column the one corresponding to the nonbasic variable with the smallest subscript.
2. Break all ties for a choice of pivotal row by picking that one of the tying rows corresponding to the basic variable with the smallest subscript.

Note that step 1 is different from our choice of pivotal column, that of picking the most negative among the cost coefficients. The negative cost coefficient that corresponds to the nonbasic variable with smallest subscript could be any one of the possible choices. If only step 2 is implemented in the rule, the cycle in Beale's example is not broken. Both steps are necessary to make cycling impossible, as shown by Bland in his proof.

A final method for breaking cycles is to make the choice of a tie breaker for incoming variables random. We know that some choice, such as lexicographic minimum, will break any cycle, so random choices will eventually find the choice that works. Each of these techniques will break the cycle and lead to the optimal solution in the example of Beale.

None of these ways for preventing cycling need to be implemented until a difficulty occurs. If a problem runs into prolonged *stalling*, i.e., the objective value remains the same for many iterations, then one of these preventive methods may be used to avoid any possible cycle. Thus, the usual rules for choosing a pivot can be maintained most of the time, even on problems that are highly degenerate.

---

[2] Bland, R. G., New Finite Pivoting Rules for the Simplex Method, *Mathematics of Operations Research* 2, (May 1977) 103–107.

4.9 FUNDAMENTAL THEOREM OF LINEAR PROGRAMMING

The Simplex Method presented in this chapter covers all linear programs. The outstanding feature is that every case terminates in a finite number of steps. This observation is summarized in the fundamental theorem, 4–3.

After a slack variable is added to each inequality constraint to reduce the inequality to equality, every linear program may be written as:

Maximize a linear function of **X** subject to

$$\mathbf{AX} = \mathbf{C}, \mathbf{X} > 0, \tag{1}$$

where **A** is an m by n coefficient matrix, **X** is an n by 1 column vector of unknowns, and **C** is an m by 1 column vector of constants.

The nonnegativity constraint means that all variables x_j, $j = 1, 2, \ldots, n$, in vector **X** are nonnegative. If all slack variables are given a zero coefficient in the objective function, then the objective may be written as **DX** where **D** is a 1 by n row vector made up of cost coefficients. Finding a solution to system (1) is known as the *Feasibility Problem*, since all components of **X** must be nonnegative. If the nonnegativity constraint is dropped then system (1) is the linear algebra problem of solving m equations in n unknowns.

Artificial variables may be added to (1) if a feasible basis is not available from the slack variables. Problems requiring artificial variables are divided into two linear programs corresponding to phase one and phase two as described in section 4.4. Phase one is the linear program (L.P.) that minimizes the sum of the artificial variables. Either this minimum is zero or the L.P. is inconsistent. The lower bound on the sum of nonnegative artificial variables is zero. If the minimum of this sum is also zero, then each artificial variable must equal zero, and $\overline{M} = M$ or the artificial objective reduces to the objective of the original problem. The corresponding tableau contains a B.F.S. to the original problem, so phase one is complete. The other possible outcome is that the sum of the artificial variables is nonzero in the final tableau of phase one. This will happen when the bottom objective row contains all nonnegative numbers, and at least one positive artificial variable remains in the basis. The existence of a nonzero artificial variable in the final basis means there are no feasible solutions to the original problem and system (1) is inconsistent.

If phase one is successful in finding a B.F.S. to (1), then the pivoting automatically continues in phase two. As we have seen, only two possible outcomes exist when phase two is complete. Either an optimal B.F.S. has been found, or there is a pivotal column with no possible pivotal row. In the latter case the objective function is unbounded over the feasible region.

Now it is easy to prove that all linear programs, whether phase one or phase two, are completed in a finite number of steps.

THEOREM 4–3. *The Fundamental Theorem of Linear Programming. For every L.P. the Simplex Method must reach one of the following three conclusions in a* *finite number of steps: (1) The L.P. is inconsistent; (2) the L.P. objective function is unbounded over the feasible region; or (3) over its feasible region the L.P. has a basic solution which optimizes the objective function.*

PROOF: Consider system (1) with artificial variables added if necessary to complete an initial B.F.S. Let n equal the total number of variables. Since in general one slack or artificial variable was added to each constraint, it is necessary that $n > m$. Choose any m of the n variables to be basic and denote the basic set of variables by $\mathbf{X_B}$. Let $\mathbf{B}$ be the matrix composed of the m columns of $\mathbf{A}$ corresponding to the basic variables. After setting the nonbasic variables equal to zero, system (1) reduces to

$$\mathbf{BX_B} = \mathbf{C}, \tag{2}$$

which has a unique solution whenever $\mathbf{B}$ is of rank m. The total number of possible choices for $\mathbf{B}$ is the number of ways m variables can be picked from n variables, or

$$\frac{n!}{m! \ (n - m)!}.$$

The Simplex Method generates only those $\mathbf{B}$ from the total number that have rank m and lead to a B.F.S. Also the objective maximum in each iteration cannot be less than that of the previous iteration. This was guaranteed by the choice of a pivotal column from among the negative coefficients in the objective row. As long as cycling is avoided by one of the methods discussed in section 4.8, the Simplex Method generates a distinct $\mathbf{B}$ in each iteration. Since the objective function is monotonic and the distinct bases come from a finite number, a B.F.S. with the optimum objective value must be reached in a finite number of steps whether in the program of phase one or the program of phase two. Finally, by theorem 4–1, no nonbasic solution to system (1) can have an objective value greater than that of the optimal B.F.S., so the proof is complete with the interpretations of the final tableau already given. ■

4.10   EFFICIENCY OF THE SIMPLEX METHOD

It is possible to create a linear program in which the Simplex Method will generate every feasible basis or every vertex of the feasible region. V. Klee and G.J. Minty (1972) gave the following example of a set of L.P.s with n variables for which the Simplex Method requires $2^n - 1$ iterations to reach a solution:

$$\max \quad \sum_{j=1}^{n} 10^{n-j} x_j$$

such that

$$(2\sum_{j=1}^{i-1} 10^{i-j} x_j) + x_i \leq 100^{i-1}, \quad i = 1, 2, \ldots, n$$

$$x_j \geq 0, \qquad j = 1, 2, \ldots, n.$$

When solved by the Simplex Method, even a small problem from this set with $n = 20$ would need over a million iterations and be unreasonably slow for the computer to solve. An algorithm is usually considered unsatisfactory if the number of iterations grows exponentially with the size of the problem. This exponential upper bound, $2^n - 1$, is a worst case bound on the number of Simplex iterations. A more valuable estimate of the efficiency of the Simplex Method is found in the average number of iterations required for solving a "typical" problem. Experience has shown that this average is a very small fraction of the possible number of bases.

G. Dantzig notes in his book (1963) that L.P.s with $m < 50$ constraints and $n < 200$ variables are usually solved by the Simplex Method in less than $3m/2$ iterations and rarely need $3m$ iterations. Similar results were found by H.W. Kuhn (1963) in a study of problems generated randomly by Monte Carlo simulation. He found that problems with 50 variables were solved on the average by $2m$ iterations, where m is the number of constraints. Recent experience with much larger problems agrees with the average number of iterations reported by Dantzig and Kuhn. Theoretic work by K. Borgwardt and S. Smale, (1982) indicates that the occurrence of problems like that of Klee and Minty, which don't share this average behavior, is so rare as to be negligible.

An example of "typical" behavior is the bomb problem discussed in section 4.4 that used four pivots and created five bases in finding the maximum explosive force. The three equations in seven variables have a total of 35 possible choices for **B**, but the Simplex Method needed to examine only five bases. The Simplex Method has been a practical way to solve L.P.s in a natural setting since the advent of high speed computers.

A recent alternative to the Simplex Method is the Russian algorithm given by L.G. Khachian[3] in 1979. Khachian's article described a significant breakthrough in mathematical theory, because it included a proof that a maximum number of iterations required by his algorithm to reach a solution is only a polynomial in the amount of problem data. The original article stated a method for solving a system of linear inequalities with integer coefficients. The method recursively creates a sequence of shrinking ellipsoids such that each contains all the feasible solutions of its predecessors. After a polynomial number of these ellipsoids have been found, either the center of one of the ellipsoids is a solution to the linear inequalities or no solution exists. The Khachian method must be adapted to L.P.s and different bounds have been found depending upon the adaptation. G. G. Brown and A. R. Washburn give an upper bound on the number of iterations as

[3] Khachian, L. G., A Polynomial Algorithm for Linear Programming (in Russian), *Doklady Akademii Nauk* USSR 244(5), 1093–1096 (1979). English translation: Soviet Mathematics Doklady 20, 191–194.

$$16 (n + m)^2 L$$

where n is the original number of L.P. variables, m is the number of constraints, and L stands for the total number of binary bits needed to store the problem coefficients in a computer. This bound is polynominal in the given data of an L.P. problem. A polynomial bound on the number of iterations is highly desirable theoretically, because a polynomial function increases at a much slower rate than an exponential function.

Is the Russian algorithm practical for solving every day L.P.s? While the number of iterations required by the Simplex Method is normally a small fraction of its upper bound, the opposite is true in the Russian case, which usually uses a number of iterations that is a significant fraction of its upper bound. E. H. McCall[4] reports a test problem of 355 constraints that was solved in 4.5 minutes by a Simplex procedure. He estimates that the same problem on the same computer would take almost 60,000 years when solved by Khachian's method. The Russian approach with its polynominal upper bound is a theoretical gem, but it is not likely to replace the Simplex Method for solving our every day linear programs.

Two alternatives to the Simplex Method will be discussed later in the text. One, known as the Revised Simplex Method, is presented in Chapter 7. It requires the same number of iterations as the Simplex Method. The other alternative is the Primal - Dual Algorithm that is given in Chapter 8. In general, this algorithm uses fewer iterations to reach a final tableau than the Simplex method. It is very efficient if the entire tableau can be kept intact in the computer's fast memory. All of these techniques contribute to the present knowledge and the ability to solve linear programs.

PROBLEMS

The following problems should be done by pivoting the condensed tableau on a computer. Programming aids are given in Chapter 5. At first the pivoting may be done one tableau at a time by telling the machine the pivot position with an input command. Eventually a completely automatic program should be compiled so that the machine will find the pivot position and continue to iterate until the final tableau is reached. Only the final tableau need be printed out. Add artificial variables where necessary and note the cases of unboundedness, inconsistency, or degeneracy.

4–19. Find nonnegative numbers x_1, x_2, x_3 and the maximum value of M for

$$x_1 + 2x_2 - x_3 \leq 9$$
$$2x_1 - x_2 + 2x_3 = 4$$
$$-x_1 + 2x_2 + 2x_3 \geq 5$$
$$2x_1 + 4x_2 + x_3 = M.$$

[4] McCall, E. H., Soviet Algorithm Not Really a "Breakthrough," *Computerworld* (Feb. 1980) 33–36.

4-20. Find nonnegative numbers x_1, x_2, x_3 and the minimum value of m for

$$x_1 + 2x_2 - x_3 \leq 9$$
$$2x_1 - x_2 + 2x_3 = 4$$
$$-x_1 + 2x_2 + 2x_3 \geq 5$$
$$2x_1 + 4x_2 + x_3 = m.$$

4-21. Find nonnegative numbers x_1, x_2, x_3 and the minimum value of m for

$$x_1 + 2x_2 - x_3 \leq 5$$
$$2x_1 - x_2 + 2x_3 = 2$$
$$-x_1 + 2x_2 + 2x_3 \geq 1$$
$$2x_1 + 4x_2 + x_3 = m.$$

4-22. Find the vector **X** and the maximum value of M that satisfies

$$x_i \geq 0 \text{ for } i = 1, 2, 3, 4$$
$$2x_1 - 4x_2 + x_3 + 3x_4 = 10$$
$$-x_1 + 4x_2 + 2x_3 - x_4 = 8$$
$$2x_1 + 3x_2 + x_3 + 2x_4 = M.$$

4-23. Find nonnegative numbers x_1, x_2, and the maximum M that satisfy

$$-3x_1 + 4x_2 \leq 0$$
$$x_1 - 2x_2 \leq 5$$
$$x_1 - x_2 \leq 6$$
$$2x_1 - x_2 \leq 14$$
$$5x_1 - x_2 \leq 39.5$$
$$20x_1 + 7x_2 \leq 202$$
$$8x_1 + 5x_2 = M.$$

Is it always best to choose the pivotal column by the most negative number in the bottom row? Answer by working this problem under both possible choices for the initial pivot. Does the degeneracy in the initial tableau affect the final result? Check your results graphically.

4-24. Run Examples 2 and 3 from Chapter 2.

4-25. Maximize $3x_1 + 3x_2$ subject to

$$x_1 \geq 0, x_2 \geq 0$$
$$3x_2 - x_1 \geq 3$$
$$x_2 - 2x_1 \leq 2.$$

4-26. Maximize $3x_2 - 2x_1$ subject to

$$x_1 \geq 0, \quad x_2 \geq 0$$
$$x_2 - 2x_1 \geq 2$$
$$3x_2 - x_1 \leq 3.$$

4-27. Maximize $4x_2 - 3x_1$ subject to

$$x_1 \geq 0, \quad x_2 \geq 0$$
$$-x_1 + 5x_2 \leq 25$$
$$4x_1 - x_2 \leq 14$$
$$x_1 + 4x_2 \geq 12.$$

4-28. Run problem 2–13 and find Mrs. Coffman's best buy.

4-29. Maximize $x_1 - x_2 + x_3$ subject to

$$x_i \geq 0, i = 1, 2, 3$$
$$-x_1 + x_2 - 2x_3 \leq 5$$
$$2x_1 - 3x_2 + x_3 \leq 3$$
$$2x_1 - 5x_2 + 6x_3 \leq 5.$$

4-30. Find vector **X** and the maximum value of M such that

$$x_i \geq 0, i = 1, 2, 3$$
$$-x_1 + 4x_2 - 2x_3 \leq 5$$
$$2x_1 - 3x_2 + x_3 \leq 3$$
$$2x_1 - 5x_2 + 6x_3 \leq 5$$
$$x_1 + x_2 + x_3 = M.$$

4–31. Find vector **X** and the maximum value of M such that

$$x_i \geq 0, \ i = 1, 2, 3, 4, 5$$
$$-2x_2 + x_3 + x_4 - x_5 \leq 0$$
$$2x_1 \qquad -2x_3 + x_4 - x_5 \leq 0$$
$$-x_1 + 2x_2 \qquad + x_4 - x_5 \leq 0$$
$$x_1 + x_2 + x_3 \qquad = 1$$
$$x_1 + x_2 - x_3 + x_4 - x_5 = M.$$

4–32. Find both the maximum and the minimum values of $x_1 + x_2 + x_3$ subject to

$$x_i \geq 0, \ i = 1, 2, 3$$
$$2x_1 - x_2 + 3x_3 = 9$$
$$5x_1 + 4x_2 - 2x_3 = 7.$$

4–33. Find both the maximum and the minimum values of $x_2 + x_3 - x_1$ subject to the constraints given in problem 4–32.

4–34. Find nonnegative numbers x_1, x_2, and the minimum value of m such that

$$7x_1 + 2x_2 \geq 14$$
$$2x_1 + x_2 \geq 6$$
$$x_1 + x_2 \geq 5$$
$$3x_1 + 2x_2 = m.$$

4–35. Run Example 1, Chapter 2, and check the graphical solution.

4–36. Run problem 2–15 and solve the sales problem of Mr. Wise.

4–37. Run problem 2–16.

4–38. Find $x_1 \geq 0$ and $x_2 \geq 0$ such that

$$3x_1 + 4x_2 \leq 12$$
$$2x_1 - 3x_2 \geq 10$$
$$3x_1 + 5x_2 = M, \text{ a maximum}$$

4–39. Determine $x_1 \geq 0$ and $x_2 \geq 0$ such that

$$9x_1 - 6x_2 \geq 3$$
$$8x_1 - 8x_2 \leq 16$$
$$2x_1 + 2x_2 = M, \text{ a maximum.}$$

4-40. Find vector **X** and the minimum value of m subject to

$$x_i \geq 0, \, i = 1, 2, 3,$$
$$8x_1 + 2x_2 + x_3 \geq 3$$
$$3x_1 + 6x_2 + 4x_3 \geq 4$$
$$4x_1 + x_2 + 5x_3 \geq 1$$
$$x_1 + 5x_2 + 2x_3 \geq 7$$
$$7x_1 + 3x_2 + 8x_3 = m.$$

4-41. Find vector **X** and the maximum value of M subject to

$$x_i \geq 0, \, i = 1, 2, 3$$
$$3x_1 - 2x_2 + x_3 \leq 5$$
$$2x_1 + x_2 - x_3 \leq 3$$
$$9x_1 - 6x_2 + 3x_3 \leq 15$$
$$12x_1 - 8x_2 + 4x_3 = M.$$

4-42. Find the maximum value of M such that

$$x_1 \geq 0, \quad x_2 \geq 0$$
$$3x_1 - 2x_2 \geq 1$$
$$x_1 - x_2 \leq 2$$
$$x_1 + x_2 = M.$$

4-43. Find vector **X** and the maximum value of M such that

$$x_i \geq 0, \, i = 1, 2, 3, 4$$
$$x_1 + 3x_2 + x_4 \leq 4$$
$$2x_1 + x_2 \leq 3$$
$$x_2 + 4x_3 + x_4 \leq 3$$
$$2x_1 + 4x_2 + x_3 + x_4 + 10 = M.$$

Hint: The value of M in the initial tableau is not zero.

4-44. Eure's Department Store must reorder six popular items. The order will consist of at least a 2 month's supply but not more than a 4 month's supply of the items. The following table contains the necessary information. Costs and prices are given in dollars and storage space is given in cubic feet per 100 items.

ITEM	COSTS	SELLING PRICE	SALES PER MONTH	STORAGE SPACE
A	1.95	2.29	800	25
B	2.49	2.99	650	30
C	5.85	6.99	300	50
D	2.60	2.99	750	20
E	9.90	12.49	250	72
F	.79	.99	900	10

The current resources of Eure's are $50,000 in cash for investment and 3000 cubic feet of storage space. How many of each of the six items should they order to earn the maximum profit neglecting the overhead expenses? Suppose increases are imminent in the wholesale costs. If you were the manager of Eure's and wished to increase the possible profit from this order, would you increase the investment capital by borrowing money or would you add more storage space?

4–45. Four types of precision rings are turned out in a machine shop on a lathe, a grinder, and a polisher. The following table gives the times in hours per batch required by each ring on each machine. Also listed is the profit in dollars per batch and the maximum number of hours available on each machine per week.

RING	LATHE	GRINDER	POLISHER	PROFIT
A	8.5	1.5	4.5	90
B	3.0	3.25	8.5	70
C	4.25	4.0	1.75	80
D	2.75	2.75	5.25	60
max. hrs.	48	40	60	

Determine the production schedule and maximum profit for the week. Chop your answers to the tenth of a batch.

4–46. If extra help is available in the machine shop of problem 4–45 and the grinder can be put to work for an additional 32 hours, then what should the production schedule and profit be? How much of the additional 32 hours is actually used on the grinder?

4–47. Farmer McPherson is considering five possible crops for his 100 acres of tillable land. He has $45,000 in credit at the local bank. For this year's growing season Mr. McPherson can count on at most 2000 man-hours of labor and 1200 hours of tractor time. He pays $9.25 per hour for labor plus $13 per hour for the tractor. The following table gives the data in terms of hours per acre or dollars per acre for each crop.

CROP	LABOR TIME	TRACTOR TIME	OTHER EXPENSES	PROFIT
barley	16	8	$90	$90
corn	20	16	100	140
beets	48	24	200	160
navy beans	40	24	75	80
potatoes	56	36	200	175

The profit is estimated on an average growing season. The other expenses include the seeds, young plants, fertilizers, insecticides, and fungicides. Mr. McPherson allows $50 per acre to plant clover on the acres not used by his crops. What should be planted and how much is the expected profit?

4–48. Suppose the market changes and the expected profit column in problem 4–47 reads respectively $80, $95, $160, $75, $185. Then what should his planting program and total profit be? In this case is he wasting any of his labor time or tractor time? Has he used all of his credit at the bank?

4–49. Bill Zippo is tuning up his car for a 500 mile race. He will need to refill his 22 gallon tank at least once. Bill uses a base gas with an added high octane mix. The mechanic assures him that his engine can use a ratio of no more than 1 gal. of the high octane mix to 2 gals. of the base gas. By trial Bill discovers that his greatest speed is attained with at least 1 gal. of high octane mix to 3 gals. of base gas. His mileage depends upon the amount of added high octane mix. If no high octane is added, then Bill's race car will average only 7 miles per gal., but this mileage will increase by 4/11 miles per gal. for each added gal. of the high octane mix. The base gas costs $1.29 per gal. and the high octane mix costs $2.39 per gal. How many gallons of each type of gas should Bill purchase in order to finish the race at maximum speed with the minimum possible fuel cost?

4–50. In problem 4–49, suppose Bill Zippo is able to tune his engine so that it will average 8 miles per gal. on the base gas. Then what mixture of gas should he buy and what is his minimum fuel cost? Check your solution graphically.

4–51. There are 1000 midshipmen due to graduate from the United States Naval Academy and receive their first duty assignment. The members of the class will each be given one of five possible assignments according to the following conditions. The new officers being assigned to the fleet must be at least one half as many as those that are going to flight school at Pensacola, Florida. No more than 250 openings exist at the nuclear power school and at the graduate school in Monterey, California combined. The number of officers to be assigned to the Marine Corps at Quantico, Virginia must be at least one and a half times the number going to Monterey, California. The number of graduates to be sent to Pensacola must be at least equal to the number going to Monterey plus twice the number going to nuclear power school. The sum of the officers going to nuclear power school and to the marines must be at least three quarters of the number going to Pensacola. Twice the number assigned to Pensacola, minus one half the number assigned to nuclear power school, minus three times the number assigned to Monterey, plus one half the number assigned to the fleet, minus twice the number assigned to Quantico is at least nine fortieths of the graduating class. Unfortunately not all of the graduates can have their first choice among the duty assignments. Experience has shown that the following numbers are happy with their assignment: 85% of those getting Pensacola, 80% of those getting the nuclear power school, 95% of those getting Monterey, 50% of those getting the fleet, and 85% of those getting the marines. How would you assign the graduating class to satisfy all of the given conditions and create the most happiness?

4–52. Determine whether the following L.P. has a solution:

$$x_i \geq 0, \ i = 1,2,3$$
$$x_1 - 2x_2 + x_3 \leq 9$$
$$3x_1 + x_2 - 4x_3 \leq 3$$
$$-3x_1 + x_2 - 3x_3 = M, \text{ a maximum.}$$

4–53. Show that the following L.P. cycles if all ties for a pivotal row are broken by removing the basic variable with the lowest subscript:

$$x_i \geq 0, \ i = 1,2,3,4$$
$$0.5x_1 - 5.5x_2 - 2.5x_3 + 9x_4 \leq 0$$
$$0.5x_1 - 1.5x_2 - 0.5x_3 + x_4 \leq 0$$
$$x_1 \leq 1$$
$$10x_1 - 57x_2 - 9x_3 - 24x_4 = M, \text{ a maximum.}$$

4–54. Break the cycle in problem 4–53 and find the optimum solution.

4–55. Show that the following L.P. does not cycle if an artificial variable is used for the equality constraint:

$$x_i \geq 0, \ i = 1,2,3,4$$

$$0.5x_1 - 5.5x_2 - 2.5x_3 + \quad 9x_4 \leq 0$$
$$0.5x_1 - 1.5x_2 - 0.5x_3 + \quad x_4 \leq 0$$
$$x_1 + \quad x_2 + \quad x_3 + \quad x_4 = 1$$
$$x_1 - \quad 7x_2 - \quad x_3 - \quad 2x_4 = M, \text{ a maximum.}$$

What is the optimum solution?

4–56. Replace the equality constraint in problem 4–55 with a type I inequality. Show that a cycle of order 6 occurs if all ties for a pivotal row are broken by removing the basic variable with the lowest subscript. Break the cycle and find an optimum solution. Is the optimum solution the same as that for problem 4–55? Are there alternate optima?

4–57. Solve:

$$x_i \geq 0, \; i = 1,2,3$$
$$x_1 - 2x_2 - 3x_3 = 1$$
$$x_1 - x_2 - 2x_3 = 0$$
$$x_1 + x_2 + x_3 = M, \text{ a maximum.}$$

Is the sum of the artificial variables equal to zero?

4–58. Solve:

$$x_i \geq 0, \; i = 1,2,3,4$$
$$x_1 - x_2 - x_3 - 2x_4 = 2$$
$$x_2 - 2x_3 + x_4 = 0$$
$$2x_1 - x_2 - 4x_3 - 3x_4 = 4$$
$$x_1 + x_2 - 4x_3 - x_4 = M, \text{ a maximum.}$$

Is one constraint redundant? Is the sum of the artificial variables equal to zero even though one remains in the basis? Are the constraints consistent?

4–59. Solve the following L.P. and count the number of pivots required:

$$x_i \geq 0, \; i = 1,2,3$$
$$x_1 \qquad\qquad \leq 1$$
$$20x_1 + x_2 \qquad \leq 100$$
$$200x_1 + 20x_2 + x_3 \leq 10{,}000$$
$$100x_1 + 10x_2 + x_3 = M, \text{ a maximum.}$$

4–60. Solve the following L.P. Does the number of required pivots agree with the Klee and Minty formula, $2^n - 1$?

$$x_j \geq 0, \qquad\qquad j = 1, 2, 3, 4$$
$$x_1 \qquad\qquad\qquad\qquad \leq 1$$
$$20x_1 + \quad x_2 \qquad\qquad\quad \leq 100$$
$$200x_1 + \; 20x_2 + \quad x_3 \qquad \leq 10{,}000$$
$$2000x_1 + 200x_2 + 20x_3 + x_4 \leq 1000000$$
$$1000x_1 + 100x_2 + 10x_3 + x_4 = M, \text{ a maximum.}$$

4–61. Solve the Klee and Minty problem for $n = 5$ and note the required number of pivots.

CHAPTER 5

Computer Subroutines in BASIC and FORTRAN

5.1   INTRODUCTION

Practical problems in linear programming involve many variables leading to large tableaux which require the computation to be carried out on a high speed computer. Even then, special techniques are used to decompose the tableau into smaller pieces so that only a part of the problem is worked on at a time. Also large tableaux are stored in the computer by entering only nonzero numbers with a flag to indicate their matrix positions. In this chapter it is assumed that the entire tableau is held in fast memory so that the algorithms developed in Chapter 4 can be applied. These iterative algorithms translate into computer programs that do the same job as done by hand for small problems. A computer program is a set of instructions, in a language understood by the computer, that tells the computer the sequence of calculations to be performed.

      The two most widely used languages for communication with computers are BASIC and FORTRAN. Each of these languages has undergone numerous revisions to make the communication more powerful or more convenient. Here we will present in both BASIC and FORTRAN the subroutines that do our pivoting. The important ingredient is the sequential logic in each subroutine. When this logic is clear, the following programs should easily adapt to any version of our two languages or to any other language. The logic is the same. Only the format differs from language to language. It's like speaking in different dialects.

The given subroutines must be incorporated into a main program. The main program controls the input and output of data as well as the calls to the various subroutines to guide the flow of the computation. Users should write their own main program to handle data in their way on their machine.

Linear programming problems can be solved with these routines on any machine, from the small personal computer to the large mainframe. Especially convenient is a *time sharing* system where many terminals are connected to a combination of central processors. These systems can handle hundreds of users simultaneously so that each user feels as if he or she were solely running the computer. A communications computer cycles the information in and out of the main processors that run continuously, doing a small portion of each user's computations at a time. The advantage of a "hands on" terminal is the immediate interaction between user and machine as in running a personal computer. This interaction is important when the user types in a pivot position and the response is the next pivoted tableau. Changes can be made at the keyboard and the progress can be readily observed.

5.2 PIVOTING THE EXTENDED TABLEAU

First we will present a subprogram in BASIC that pivots the extended tableau. Following the notation of Chapter 4, variable $Y(I,J)$ will stand for the tableau entry in the ith row and j^{th} column. The size of the tableau will be indicated by variables M, the number of rows, and N, the number of columns. The main program will read into the computer the numbers M, N and the tableau entries $Y(I, J)$, $I = 1,2,\ldots, M$, and $J = 1,2,\ldots, N$. This is easily done with READ and DATA statements as follows:

```
300  READ M,N
380  FOR I=1 TO M
390  FOR J=1 TO N
400  READ Y( I, J )
410  NEXT J
420  NEXT I
```

The DATA statements contain the corresponding numbers in order where the tableau is given row by row.

The main program also determines the pivot position P,Q, where P is the pivotal row and Q is the pivotal column. Then a call statement such as

```
450 CALL PIVEX( Y,M,N,P,Q )
```

will cause the following subprogram to carry out one pivoting iteration.

```
490  SUB PIVEX ( Y( , ),M,N,P,Q )
500  LET A=Y( P,Q )
504  FOR J=1 TO N
510  LET Y( P,J )=Y( P,J )/A
520  NEXT J
530  FOR I=1 TO M
540  IF I=P THEN 580
550  IF Y( I,Q )=0 THEN 580
555  LET B=Y( I,Q )
560  FOR J=1 TO N
570  LET Y( I,J )=Y( I,J )-B*Y( P,J )
580  NEXT J
590  NEXT I
600  END SUB
```

The result of subprogram PIVEX, which is returned to the main program, is the pivoted tableau Y with a unit column vector in the pivotal column. The brackets in the subprogram mark the loops so that the logical sequence is clear. In the first loop, statement 510 leaves a one in the pivot position when $J = Q$. The pair of nested loops produce the zeros down the pivotal column except at the pivot position. The IF statement in 540 skips the pivotal row. The inner loop carries out the Gauss-Jordan elimination across the ith row while the outer loop merely advances the computation from row to row. Note that the multiple of the pivotal row, B, is stored so that it will not change during the elimination in 570.

An alternative to the subprogram in BASIC is to replace the call statement with GOSUB 500. If GOSUB with the appropriate line number is used, then the END SUB statement is replaced with RETURN. The GOSUB is simple to use, but it does not allow the internal variables of the subroutine to be independent of corresponding variables in the main program.

PIVEX has a number of applications. It will solve a set of n linear equations in n unknowns provided the rank of the coefficient matrix is equal to n, theorem 1–9, corollary 1. If a pivot is not zero or is so small that the computation becomes unstable, then the augmented matrix can be pivoted down its main diagonal. After pivoting down the main diagonal of this n by $n+1$ tableau, the solution vector appears in the right-hand column and the remainder of the tableau is the identity matrix. See section 1.6 for the appropriate theory and problem 13 in this chapter for an improvement in the technique.

Another application of PIVEX is found in the inversion of a nonsingular matrix. Computing the inverse of a nonsingular matrix by pivoting was illustrated in section 1.7. The data for this computation is an n by $2n$ tableau. The first n columns of the tableau are made up of the columns of n by n matrix **A**.

The second n columns in the tableau contain $\mathbf{I}''$, the n by n identity matrix. In its simplest version the main program will call PIVEX for each pivot down the main diagonal of $\mathbf{A}$. After n iterations the result is the identity, $\mathbf{I}''$, in the first n columns, and the inverse, $\mathbf{A}^{-1}$, in the second n columns. Only the second n columns need to be output to see the inverse matrix.

PIVEX can also be used to pivot the extended tableau of a linear program to produce a new basic feasible solution. However, we will solve linear programs by pivoting the condensed tableau.

The same subroutine is given below in the FORTRAN language. The only difference in notation is that the pivot position is designated IP, IQ to make P and Q integer variables. The DIMENSION statement reserves space for tableau Y. Allow ample space for the size of the problems to be run. A similar DIMENSION statement should appear in the main program reserving the same space. This way of reserving space for a matrix or a vector is also used in the main program in the BASIC language. The steps in the following FORTRAN program should be easy to compare with the corresponding steps in the BASIC program.

```
      SUBROUTINE PIVEX ( IP, IQ, M, N, Y )
      DIMENSION Y ( 25, 25 )
      A = Y ( IP, IQ )
      DO 10 J = 1, N
   10 Y ( IP, J ) = Y ( IP, J )/A
      DO 50 I = 1, M
      IF ( I−IP ) 30, 50, 30
   30 B = Y ( I,IQ )
      DO 40 J = 1, N
   40 Y ( I, J ) = Y( I,J )−B•Y ( IP, J )
   50 CONTINUE
      RETURN
      END
```

5.3 PIVOTING THE CONDENSED TABLEAU

To pivot the condensed tableau we introduce two new subscripted variables, K(J), J=1 to N−1, and L(I), I=1 to M−1. The K(J) stand for the subscripts of the variables out of the basis that appear across the top of our tableau. The L(I) stand for the subscripts of the variables that are in the basis and appear to the left of our tableau. The initial values of K(J) and L(I) are read in by the main program along with the initial tableau.

After a pivot position is determined, the following subprogram in BASIC will carry out the pivoting calculations.

```
1140   SUB PIVCO ( Y( , ),K( ), L( ),M,N,P,Q )
1150   LET R=1/Y( P,Q )
1160   FOR J=1 TO N
1170   Y( P,J )=R*Y( P,J )
1180   NEXT J
1190   Y( P,Q )=R
1200   FOR I=1 TO M
1210   IF I=P THEN 1270
1220   FOR J=1 TO N
1230   IF J=Q THEN 1250
1240   Y( I,J )=Y( I,J )−Y( I,Q )*Y( P,J )
1250   NEXT J
1260   Y( I,Q )=−R*Y( I,Q )
1270   NEXT I
1280   LET X=K( Q )
1290   LET K( Q )=L( P )
1300   LET L( P )=X
1320   END SUB
```

The five steps of the Condensed Tableau Pivoting Algorithm from Chapter 4 can be seen in this subprogram. Statement 1150 merely stores the reciprocal of the pivot. The first loop marked by brackets carries out step 2 of our algorithm. It multiplies the pivotal row by the reciprocal of the pivot. Statement 1190 is step 1 of our algorithm, replacing the pivot with its reciprocal. The inner loop of the nested pair of loops carries out step 4 in the algorithm. The two IF statements skip both the pivotal row and the pivotal column before carrying out algorithm step 4. Step 4 applies only to numbers not in the pivotal row and not in the pivotal column. The inner loop computes entries across the ith row while the outer loop moves the computation from row to row down the tableau. Statement 1260 corresponds to algorithm step 3, multiplying the numbers in the pivotal column by the negative reciprocal of the pivot. Again the pivot has been skipped before carrying out step 3. Finally, step 5 is carried out by statements 1280 through 1300. Note that the subscript K(Q) is temporarily stored in location X so that it will not be destroyed when replaced by subscript L(P).

After a call to subprogram PIVCO, if we wish to see the pivoted tableau with the new B.F.S., the main program must print out the result. A set of loops in BASIC that will print the tableau follows:

```
800   FOR J=0 TO N
810   PRINT K( J );
820   NEXT J
830   PRINT
840   PRINT
```

```
850   FOR I=1 TO M
860   PRINT L( I );
870   FOR J=1 TO N
880   PRINT Y( I,J );
890   NEXT J
900   PRINT
910   NEXT I
920   PRINT
```

K(0) and K(N) are set equal to zero and are included to retain the spacing across the top of the tableau. Likewise, L(I) is given some positive number for the objective rows.

Subroutine PIVCO is given below in FORTRAN.

```
      SUBROUTINE PIVCO ( Y, K, L, M, N, IP, IQ )
      DIMENSION Y ( 30, 30 ), K( 30 ), L( 30 )
      R=1 ./Y ( IP,IQ )
      DO 10 J=1, N
10    Y ( IP,J )=R*Y( IP,J )
      Y( IP,IQ )=R
      DO 50 I=1, M
      IF ( I−IP ) 20, 50, 20
20    DO 40 J=1, N
      IF ( J−IQ ) 30, 40, 30
30    Y ( I,J )=Y ( I,J )−Y ( I,IQ ) * Y ( IP,J )
40    CONTINUE
      Y( I,IQ )=−R*Y( I,IQ )
50    CONTINUE
      X=K( IQ )
      K( IQ )=L( IP )
      L( IP=X )
      RETURN
      END
```

5.4 AUTOMATIC PIVOTING SUBROUTINES

We wish to have the computer find the pivot position and automatically continue to pivot until the final tableau is reached. The main program need print out only the final tableau at the conclusion of a problem. Three additional subroutines make the whole process automatic.

The first of these is our column search which determines a pivotal column if any is available. If there is no possible pivotal column, then the following subroutine returns Q= −1 to the main program.

```
940   SUB COL ( Y( , ),K( ),M,N,Q )
950   LET Q= −1
```

```
960    LET B=0
970    FOR J=1 TO N−1
980    IF K( J )< 0 THEN 1030
990    IF Y( M,J ) > = 0 THEN 1030
1000   IF ( Y( M,J )−B )>= 0 THEN 1030
1010   LET B=Y( M,J )
1020   LET Q=J
1030   NEXT J
1040   END SUB
```

The bracketed loop runs through the first $n-1$ columns because we never pivot in the last column. Notice that the first IF statement prevents any artificial variable from being brought back into the basis. If K(J) is negative, that column is skipped and never used again in choosing a pivot. Statement 990 skips any positive or zero entries in the objective row. Thus, only columns with a negative objective coefficient are considered. Statements 1000 and 1010 ensure that the negative number of greatest absolute value is chosen in the objective row. Then statement 1020 sets Q equal to the pivotal column. The main program has a pivotal column whenever Q is positive or knows there is no pivotal column when Q is negative. When Q becomes negative the final tableau has been reached and should be printed.

The FORTRAN version of COL is given next.

```
       SUBROUTINE COL ( Y, K, M, N, IQ )
       DIMENSION Y( 30, 30 ), K( 30 )
       IQ = −1
       B = 0
       NN = N − 1
       DO 30 J = 1, NN
       IF ( K( J ) ) 30, 10, 10
10     IF ( Y( M,J ) ) 20, 30, 30
20     IF ( Y( M,J ) −B ) 25, 30, 30
25     B = Y ( M, J )
       IQ = J
30     CONTINUE
       RETURN
       END
```

The second subroutine needed by the main program is ROW. After finding a pivotal column, the main program calls ROW to search for a pivotal row if one is available. If none is available, BASIC program ROW returns P = −1 to the main program.

```
1045   SUB ROW ( Y( , ),M,N,P,Q )
1050   LET P=−1
1060   LET S=1E19
```

```
┌1070   FOR I=1 TO M−1
│1080   IF Y( I,Q )<= 0 THEN 1130
│1090   LET R=Y( I,N )/Y( I,Q )
│1100   IF ( S−R )<= 0 THEN 1130
│1110   LET S=R
│1120   LET P=I
│1130   NEXT I
└1140   END SUB
```

The bracketed loop runs through the first $M - 1$ rows skipping the objective row. M must be lowered one more when there are two objective rows. The first IF statement skips any negative or zero numbers in the pivotal column. Thus, only positive θ ratios are considered in statement 1090. Statements 1100 and 1110 assure us that the smallest of the θ ratios is chosen before setting P equal to the corresponding row. If a tie for least occurs among the θ ratios, this program picks out the first of the tying rows encountered. If P is positive, the main program should call for a pivot of the tableau at position P, Q with statement

CALL PIVCO (Y,K,L,M,N,P,Q)

After pivoting, the main program again calls for the column search. Whenever P is returned negative, the main program should print the final tableau.

Subroutine ROW in FORTRAN follows:

```
      SUBROUTINE ROW ( Y, M, N, IP, IQ )
      DIMENSION Y(30, 30)
      IP = −1
      S = .1 E 19
      MM =M−1
   ┌  DO 80 I = 1, MM
   │  IF ( Y( I,IQ ) ) 80, 80, 10
   │10 R=Y ( I,N )/Y ( I, IQ )
   │  IF (S−R) 80, 80, 20
   │20 S = R
   │  IP = I
   └80 CONTINUE
      RETURN
      END
```

The third subroutine is needed to distinguish between phase one pivoting and phase two pivoting. The main program must know if any artificial variables remain in the basis. If artificial variables exist, then it must call the column search at row M. This is phase one where negative numbers are looked for in the last row. As soon as the last artificial variable leaves the basis , the main program must call the column search at row M − 1. This search begins phase two where negative numbers are sought in the first of the two objective rows. Of course if the original problem had no artificial variables, then only one objective row called row M is used. To allow for these three possible branches, LOOK is called before each column search until the artificial variables are removed.

```
1310   SUB LOOK ( L( ),M,V )
1320   FOR I=1 TO M
1330   IF L( I )<0 THEN 1370
1340   NEXT I
1350   LET V=1
1360   GO TO 1380
1370   LET V = −1
1380   END SUB
```

For convenience all of the original rows of the tableau are numbered on the left so that L(I) is defined from 1 to M. The objective rows may be given any positive number. LOOK returns V=−1 if any L(I) is negative, and it returns V=1 if all L(I) are positive. At any stage the main program knows if there are artificial variables in the basis by a call to LOOK and then a test of the sign of V.

One branch in the main program is used for problems with one objective row and no artificial variables. This branch is followed when the first call to LOOK returns V positive. A second branch carries out phase one pivoting as long as V returns negative. On this branch COL is called at row M, and ROW is called with M = M − 1 so that θ ratios are not computed in the two objective rows. The third branch corresponds to phase two pivoting and is entered when V changes from negative to positive. On this branch COL is called at row M − 1, and ROW is still called with M = M − 1.

In FORTRAN subroutine LOOK is as follows:

```
       SUBROUTINE LOOK ( L, M, V )
       DIMENSION L ( 30 )
       DO 10 I=1, M
       IF ( L( I ) ) 30, 10, 10
10     CONTINUE
       V = 1
       RETURN
30     V = −1
       RETURN
       END
```

COLUMN, ROW and LOOK are the three subroutines necessary to carry out an automatic pivoting program. Care must be taken in the main program to use the row search only over the first M − 2 rows for problems with a double objective row. We never pivot on one of the objective rows. Upon reaching the final tableau the program should print out the K(J) and L(I) in their respective positions along with the Y(I,J) for a complete interpretation of the results. Finally, it is nice to put in a counter to tell the number of times that the original tableau was pivoted.

PROBLEMS

5-1. Write a BASIC or FORTRAN program that will pivot a given tableau down its main diagonal to produce an $M \times M$ identity matrix.
Hint: The principal step is a loop that will set the pivot and call PIVEX.

```
        FORTRAN
        DO 60 K = 1, M
        IP = K
        IQ = K
        CALL PIVEX (IP, IQ, M, N, Y)
      60 CONTINUE
          BASIC
      FOR K = 1 TO M
      LET P = K
      LET Q = K
      CALL PIVEX (Y, M, N, P, Q)
      NEXT K
```

Use your BASIC or FORTRAN program to solve the following six systems of linear equations by pivoting down the main diagonal.

5-2. $3x_1 + 2x_2 - x_3 = -4$

$2x_1 - 5x_2 - 4x_3 = 0$

$8x_1 - 3x_2 - 2x_3 = 8$

5-3. $3x_1 - 4x_2 + 2x_3 = 8$

$x_1 + 3x_2 - 5x_3 = 13$

$6x_1 - 4x_2 - 7x_3 = 9$

5-4. $5x_1 - x_2 + x_3 = 12$

$x_1 + 2x_2 + 3x_3 = 15$

$4x_1 + 2x_2 - 3x_3 = 5$

5-5. $x_1 - 2x_2 - 3x_3 + 4x_4 = 5$

$-6x_1 + 7x_2 + 8x_3 + 9x_4 = 12$

$9x_1 - 8x_2 + 7x_3 - 6x_4 = 16$

$5x_1 - 4x_2 - 3x_3 - 2x_4 = 10$

5-6. $5x_1 - x_2 + x_3 = 11$

$x_1 + 2x_2 + 3x_3 = 17$

$4x_1 + 2x_2 - 3x_3 = 3$

5-7. $2x_1 - 4x_2 + 3x_3 - 5x_4 + 2x_5 = -13$

$3x_1 + 2x_2 + 5x_3 + 4x_4 + 6x_5 = -7$

$-4x_1 + 5x_2 - 6x_3 + 3x_4 - 2x_5 = 15$

$5x_1 + 3x_2 - 4x_3 - 8x_4 + 7x_5 = 9$

$-6x_1 - x_2 + 9x_3 - 3x_4 - 5x_5 = -12$

5-8. Write a BASIC or FORTRAN program that will invert a given nonsingular matrix without zeros at a pivot position.

5-9. The nth order Hilbert matrix may be defined as follows.

$$\begin{pmatrix} 1 & 1/2 & 1/3 & \cdots & 1/n \\ 1/2 & 1/3 & 1/4 & \cdots & \dfrac{1}{(n+1)} \\ 1/3 & 1/4 & 1/5 & \cdots & \dfrac{1}{(n+2)} \\ \vdots & & & & \vdots \\ 1/n & \dfrac{1}{(n+1)} & \dfrac{1}{(n+2)} & \cdots & \dfrac{1}{(2n-1)} \end{pmatrix}$$

Invert the second order Hilbert matrix by hand to be clear on the process used in section 1.7. As a check multiply the Hilbert matrix by its inverse to see if the product is the appropriate identity matrix.

5-10. Invert the third order Hilbert matrix by using your BASIC or FORTRAN program.

Hint: The matrix should be constructed by a double loop in your program. The only data is the order N. Such a set of nested loops follows.

```
           BASIC                              FORTRAN
    55   LET K=-1                       B= -1.
    60   FOR I=1 TO M                   DO 80 I=1, N
    65   LET K=K+1                      B=B+1
    70   FOR J=1 TO M                   DO 80 J=1, N
    80   LET Y(I,J)=1/(J+K)             A=J
    90   NEXT J                     80  Y(I,J) =1./(A + B)
    95   NEXT I
```

5-11. Invert the fourth order Hilbert matrix with your program.

5-12. Invert the fifth order Hilbert matrix. Notice the size of the numbers in your inverse. The determinant of the Hilbert matrix is small so the numbers in its inverse grow very rapidly with the order. Round-off error soon becomes important. You may check on the size of the error by having your machine multiply the Hilbert matrix by its inverse and seeing how close the product is to the identity matrix.

5-13. There is a serious flaw in the program of problem 5-1; it blows up if the pivot happens to be zero or close to zero. Correct this flaw by the following technique. Search down the pivotal column below the pivot for the entry of greatest absolute value. If this absolute value is greater than that of the pivot, then interchange the pivotal row with this new row before carrying out the pivoting. Thus the largest numerical value available will always be used for the pivot. In particular zero will be avoided if possible. If the pivot and all entries below it are zero, then the matrix is singular and the system of equations has no unique solution.

Hint: After setting the pivot position, the following loop will determine which row has the greatest absolute value in the pivotal column.

```
          BASIC
   135   LET X=ABS(Y(P,Q))
   140   LET L=P
   145   FOR M1=P+1 TO M
   150   IF ABS(Y(M1,Q))<=X THEN 165
   155   LET X=ABS(Y(M1,Q))
   160   LET L=M1
   165   NEXT M1


          FORTRAN
          X=ABS (Y(IP,IQ))
          L=IP
          NN=IP + 1
          DO 60 MM=NN, M
          IF (ABS (Y(MM,IQ)) −X) 60, 60, 55
   55     X=ABS (Y(MM,IQ))
          L=MM
   60     CONTINUE
```

After this loop is executed L will denote the desired row. If X is not too small and L is different from IP, then interchange row IP with row L. Finally call PIVEX and then continue to the next pivot position.

5-14. Use your revised program from problem 5-13 to solve the following system.

$$9x_2 - 2x_3 = 14$$
$$4x_1 + 5x_2 + 3x_3 = -1$$
$$6x_1 + 7x_2 + 4x_3 = 2$$

5-15. Does the following system have a unique solution?

$$x_1 + 2x_2 + 5x_3 = 21$$
$$2x_1 + 4x_2 + 3x_3 = 7$$
$$3x_1 + 6x_2 + 2x_3 = -2$$

5-16. Solve:

$$4x_2 - 7x_3 = 13$$
$$3x_1 \qquad + 5x_3 = 0$$
$$6x_1 + 11x_2 \qquad = 8.$$

5-17. Solve:

$$x_1 + \tfrac{1}{2}x_2 + \tfrac{1}{3}x_3 = 32$$
$$\tfrac{1}{2}x_1 + \tfrac{1}{3}x_2 + \tfrac{1}{4}x_3 = 22$$
$$\tfrac{1}{3}x_1 + \tfrac{1}{4}x_2 + \tfrac{1}{5}x_3 = 17.$$

5-18. Solve:

$$x_1 + \tfrac{1}{2}x_2 + \tfrac{1}{3}x_3 + \tfrac{1}{4}x_4 = 167$$
$$\tfrac{1}{2}x_1 + \tfrac{1}{3}x_2 + \tfrac{1}{4}x_3 + \tfrac{1}{5}x_4 = 125$$
$$\tfrac{1}{3}x_1 + \tfrac{1}{4}x_2 + \tfrac{1}{5}x_3 + \tfrac{1}{6}x_4 = 101$$
$$\tfrac{1}{4}x_1 + \tfrac{1}{5}x_2 + \tfrac{1}{6}x_3 + \tfrac{1}{7}x_4 = 85.$$

5-19. Write a BASIC or FORTRAN program that will pivot a given condensed tableau at a given position. Hint: The principal step is an input command that will allow the operator to type in the pivot position (IP, IQ). Then a call to PIVCO completes the job.

5-20. Write a main BASIC or FORTRAN program that will automatically pivot a given tableau until the final tableau is reached. The pivot position is found by a call to COL and, if successful, followed by a call to ROW. If artificial variables are present then LOOK is used for branching. Provide for three branches. One branch is for tableaux with a single objective row. The other two branches are for phase one and phase two pivoting of tableaux with a double objective row.

CHAPTER 6

Matrix Games and the Concept of Duality

6.1 INTRODUCTION AND DEFINITION OF A MATRIX GAME

Since duality arises so naturally in the setting of a matrix game, we will examine a little elementary game theory. The discussion will be restricted to games with a finite number of outcomes that can be solved as linear programs on a computer.

The development of a theory of game strategy began in the 1920s and continues to this day. The theory made its greatest advance under John von Neumann, who deserves credit for the extensive development as presented in his classic work, *Theory of Games and Economic Behavior*, published jointly with Oskar Morganstern.[1] In 1928 von Neumann gave the first proof of the central theorem of matrix games, the famous Minimax Theorem. At the time linear programming was not available, so his proof used the fixed point theorems of L.E.J. Brouwer and other advanced mathematics.

The connection between game theory and linear programming is not obvious, but was recognized by George Dantzig who offered an elementary

---

[1] John von Neumann and Oskar Morganstern, *Theory of Games and Economic Behavior*, 1953.

constructive proof of the Minimax Theorem in 1956.[2] We will demonstrate the connection between the two when we express the solution to a matrix game in a linear programming format.

DEFINITION 1. A *matrix game* is a two-person game described by an $m \times n$ pay-off matrix.

One person A will choose rows while the other person B chooses columns. An entry, Y_{ij} in the pay-off matrix tells the amount of money to be paid to player A by player B whenever the ith row and jth column are chosen independently by the two players.

The entries Y_{ij} in the pay-off matrix may be positive, negative, or zero. A positive pay-off represents a gain for player A while a negative pay-off is a loss for A, that is, a pay-off to player B. A zero entry is a tie or no pay-off.

A matrix game as described here is an example of a finite, zero-sum, two-person game. There are only a finite number of choices for each player. The term *zero-sum* means that the algebraic sum of the money won and lost by all players is zero. Wealth is neither created nor destroyed but merely redistributed in a zero-sum game. A slot machine is not a zero-sum game because the house takes a healthy cut before any pay-off to the players. A private game of blackjack is zero-sum because what one individual gains others lose.

In two-person game theory, we are interested in the plays that will win the most money for player A, and also the plays that are best for player B. Of course, B is looking for negative pay-offs in the matrix. He wishes to pay the least amount possible to A by picking the most negative pay-offs. The picture of duality arises in solving for the best possible strategies for each of the players.

6.2 ANALYSIS OF 2 × 2 MATRIX GAMES

A simple example of a matrix game is the matching of pennies between two persons.

Example 1. The following pay-off matrix shows that player A receives a penny when the coins match and he pays out a penny when the coins differ.

		B	
		heads	tails
A	heads	1	-1
	tails	-1	1

The problem for both A and B is how often to play heads and how often to play tails. This leads us to the idea of relative frequency or probability.

[2] George B. Dantzig, "Constructive Proof of the Min-Max Theorem," *Pacific Journal of Mathematics*, 6(1)(1956): 25.

DEFINITION 2. The *probability* of making a particular play i is a rational number x, $0 \le x \le 1$, that is the ratio of the number of times i will be played to the total number of plays of the game.

Suppose we assume A and B can choose at will their own probability of heads appearing. An arbitrary probability can be achieved by setting up a spinner on a circle. For example, to set up a probability of heads of .63 and a corresponding probability of tails of .37, divide the circle into 100 sectors with an arbitrary 63 sectors marked heads and with the other 37 sectors marked tails. If A has chosen .63 for heads, a spin of the spinner constitutes a play for A. B can set up a different probability of heads and tails with a different spinner and appropriately divided circle. A probability of zero for some choice means that choice will never be played, and a probability of one means that choice will be taken every time. The problem is to determine a pair of probabilities that is optimal for A and another pair that is optimal for B.

DEFINITION 3. A *mixed strategy* for n choices is an ordered n-tuple of probabilities, $(x_1, x_2, \ldots, x_n)$, such that $x_1 + x_2 + \cdots + x_n = 1$.

DEFINITION 4. The *ith pure strategy* is a mixed strategy with 1 in the ith place and 0 elsewhere.

In the case of the penny match, let A's mixed strategy be $(x_1, x_2) = (x, 1 - x)$ and B's mixed strategy be $(y, 1 - y)$, where $0 \le x \le 1$ and $0 \le y \le 1$.

$$\begin{array}{cc} & B \\ & \begin{array}{cc} y & 1-y \end{array} \\ \begin{array}{c} A \end{array} \begin{array}{c} x \\ 1-x \end{array} & \begin{array}{|cc|} \hline 1 & -1 \\ -1 & 1 \\ \hline \end{array} \end{array}$$

Let z be A's expectation per play, that is, the amount A can expect to gain on the average per play of the game. With probability x, A can expect to gain 1¢ under B's probability y while losing 1¢ with probability $(1 - y)$. Thus, from the first row, A can expect to gain

$$x[1y + (-1)(1 - y)].$$

Likewise with probability $(1 - x)$, A can expect to lose 1¢ with probability y and gain 1¢ with probability $(1 - y)$. A's expected gain in the second row is then

$$(1 - x)[(-1)y + 1(1 - y)].$$

The sum of these two, z, is the total expected gain of A per play.

$$z = x[\ 1y + (-1)(1 - y) \ ] + (1 - x)[\ (-1)y + 1(1 - y) \ ]$$

$$= x(2y - 1) + (1 - x)(1 - 2y)$$

$$= 4xy - 2x - 2y + 1 \qquad\qquad (1)$$

$$= 4(\ xy - \tfrac{1}{2}x - \tfrac{1}{2}y \ ) + 1$$

$$= 4(\ x - \tfrac{1}{2})(\ y - \tfrac{1}{2}). \qquad\qquad (2)$$

In the general case of an $m \times n$ pay-off matrix, we have the following definition.

DEFINITION 5. The *mathematical expectation* of player A is the double summation,

$$z = \sum_{j=1}^{m} \sum_{j=1}^{n} x_i \, Y_{ij} y_j,$$

where $(x_1, x_2, \ldots, x_m)$ and $(y_1, y_2, \ldots, y_n)$ are any mixed strategies for A and B, and Y_{ij} are the entries in the pay-off matrix.

An analogous expression with the summations reversed can be written down for the mathematical expectation of player B. However, each sum has only a finite number of terms so the order is immaterial. This means that the definition of mathematical expectation of B is identical to that of A. The mathematical expectation of either player is the amount he expects to win or pay out, on the average, per play of the game, under his strategy.

The object of the game from A's viewpoint is to make his expectation z as large as possible. A negative value of z indicates a loss for A or a gain by player B. B's objective is to minimize z so that he pays the least amount possible to A. Of course, if B can make z negative then he wins that amount on the average per play.

To see what strategy is optimal consider in equation (2) that player A chooses his probability of heads to be $x < \tfrac{1}{2}$. Then the factor $(x - \tfrac{1}{2})$ is negative. Player B, after noticing this trend, can always win by making his factor $(y - \tfrac{1}{2})$ positive, that is, by choosing $y > \tfrac{1}{2}$. On the other hand if A chooses $x > \tfrac{1}{2}$, then $(x - \tfrac{1}{2})$ is positive. Player B can again always win by making $(y - \tfrac{1}{2})$ negative, choosing in this case $y < \tfrac{1}{2}$. The only way that player A can avoid losing is by choosing $x = \tfrac{1}{2}$. A's optimal strategy is then $(\tfrac{1}{2}, \tfrac{1}{2})$. Using exactly the same reasoning, B's optimal strategy is to choose $(\tfrac{1}{2}, \tfrac{1}{2})$. Thus, the best play for both A and B is to take an independent, random toss to ensure that they play heads or tails with equal frequency. ●

The optimal mixed strategies are computed from a conservative view. That is, player A asks the question, "What is my minimum return under a given strategy?" He then tries to maximize this minimum. By choosing $x = \tfrac{1}{2}$ in the penny match game, A guarantees his expectation to be at least zero no matter what B does. Thus on the average A can do no worse than break even. Player B asks the question, "What is my maximum pay-off to A under a given strategy?" He then tries to minimize this maximum. The value B arrives at in this way is the most he must pay out regardless of what A does.

DEFINITION 6. The *value* of a matrix game *to player A* is his expectation computed under his optimal strategy no matter what B does. The game *value to player B* is B's expectation computed under his optimal strategy no matter what A does.

A's game value is found by determining the strategy that will maximize his minimum expectation, and B's game value is found by determining the strategy that will minimize his maximum expectation.

DEFINITION 7. The game is called *fair to player A* if his value is zero, and the game is *fair to player B* if his value is zero.

The value of the penny match game is $z = 0$ for both players and so the game is fair to both. This is not a mere coincidence.

Example 2. Consider a different pay-off matrix:

$$B$$

		y	$1 - y$
	x	-3	6
A			
	$1- x$	2	-5

If A and B choose strategy $(\,{}^1\!/_2,\,{}^1\!/_2\,)$ then $z = 0$ and the game seems to be fair. Let us compute $z = f(\,x,\,y\,)$ and see if the game is fair.

$$z = x[\,-3y + (\,1 - y\,)6\,] + (\,1 - x\,)[\,2y - 5(\,1 - y\,)\,]$$
$$= x(\,6 - 9y\,) + (\,1 - x\,)(\,7y - 5\,)$$
$$= -16xy + 11x + 7y - 5 \tag{3}$$
$$= -16(\,xy - {}^{11}\!/_{16}x - {}^{7}\!/_{16}y\,) - 5$$
$$= -16(\,x - {}^{7}\!/_{16}\,)(\,y - {}^{11}\!/_{16}\,) + {}^{77}\!/_{16} - 5 \tag{4}$$
$$= -16(\,x - {}^{7}\!/_{16}\,)(\,y - {}^{11}\!/_{16}\,) - {}^{3}\!/_{16}$$

Player A controls factor $(x - {}^{7}\!/_{16}\,)$ and B controls factor $(\,y - {}^{11}\!/_{16}\,)$. If either one makes the value of their factor different from zero, the other player can take the advantage. Thus the optimal strategies are ($\frac{7}{16}$, $\frac{9}{16}$) for A and ($\frac{11}{16}$, $\frac{5}{16}$) for B. The value of the game is $\frac{3}{16}$ to both players, which is not fair to player A. On the average player A will lose 3¢ every 16 plays. ●

6.3   COMPUTATION OF STRATEGIES[3]

The computation may be simplified by using calculus. Consider the locus of $z = f(x, y)$ as a surface in three dimensional space. The surfaces represented by equations (1) and (3) are hyperbolic paraboloids. The hyperbolic paraboloid is distinguished by its *saddle* shape and a point called the *saddle point*. The saddle point is the minimum of parabolas on the surface opening upward and the maximum of other parabolas on the surface opening downward. Cross sections orthogonal to the planes of these parabolas are hyperbolas. Figure 6–1 represents the hyperbolic paraboloid $z = xy$.

Relative extreme points and saddle points are found on differentiable surfaces by the following sequence of theorems. The subscripts stand for partial derivatives with respect to the variables in the subscript.

THEOREM 6–1.   *If $f(x, y)$ has continuous first partial derivatives, necessary conditions for f to have an extreme at (a, b) are $f_x = 0$ and $f_y = 0$ at (a, b).*

A set of sufficient conditions for a relative extreme is given by theorem 6–2.

THEOREM 6–2.   *If $f(x, y)$ has continuous second partial derivatives, if $f_x = f_y = 0$ at (a, b), and if $f^2_{xy} - f_{xx}f_{yy} < 0$ at (a, b), then the point (a, b) is an extreme of f. The extreme will be a maximum if $f_{xx} < 0$ and a minimum if $f_{xx} > 0$ at (a, b).*

---

[3] This section may be skipped by those not familiar with two variable calculus.

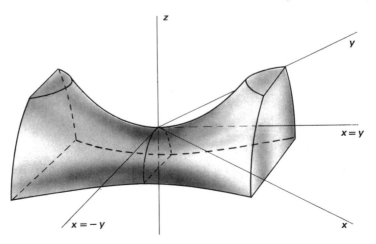

Figure 6–1

Theorem 6-3.[4] *If $f(x, y)$ has continuous second partial derivatives, if $f_x = f_y = 0$ at (a, b), and if $f^2_{xy} - f_{xx}f_{yy} > 0$ at (a, b), then the point (a, b) is a saddle point of f.*

Applying theorem 6-3 to example 1, we have

$$f(x, y) = 4xy - 2x - 2y + 1$$
$$f_x = 4y - 2 = 0, \qquad f_y = 4x - 2 = 0$$
$$y = \tfrac{1}{2}, \qquad\qquad x = \tfrac{1}{2}$$
$$f_{xx} = 0, \qquad f_{xy} = 4, \qquad f_{yy} = 0$$
$$f^2_{xy} - f_{xx}f_{yy} = 16 > 0.$$

Thus the point $(\tfrac{1}{2}, \tfrac{1}{2}, 0)$ on the surface is a saddle point. From example 2, we find

$$f(x, y) = -16xy + 11x + 7y - 5$$
$$f_x = -16y + 11 = 0, \qquad f_y = -16x + 7 = 0$$
$$y = \tfrac{11}{16}, \qquad\qquad x = \tfrac{7}{16}$$
$$f_{xx} = 0, \qquad f_{xy} = -16, \qquad f_{yy} = 0$$
$$f^2_{xy} - f_{xx}f_{yy} = 256 > 0.$$

This checks that the point $(\tfrac{7}{16}, \tfrac{11}{16}, -\tfrac{3}{16})$ is a saddle point. In both cases the coordinates of the saddle point give the optimal strategies and the value of the game. This will be true in general provided the saddle point occurs over the unit square, $0 \le x \le 1, 0 \le y \le 1$.

6.4 AN EXAMPLE OF PURE STRATEGIES

Let us look at one more example of a 2×2 matrix game.

Example 3.

		B	
		y	$1 - y$
A	x	1	2
	$1 - x$	-2	1

---

[4] For proof see D.V. Widder, *Advanced Calculus*, second edition, p. 128.

For this game A's expectation is

$$z = x[\, y + 2(\, 1 - y\,)\,] + (\, 1 - x\,)[\, -2y + (\, 1 - y\,)\,]$$
$$= x(\, 2 - y\,) + (\, 1 - x\,)(\, 1 - 3y\,)$$
$$= 2xy + x - 3y + 1$$
$$= 2(\, xy + x/2 - 3y/2\,) + 1$$
$$= 2(\, x - {}^3\!/_2\,)(\, y + {}^1\!/_2\,) + {}^3\!/_2 + 1$$
$$= 2(\, x - {}^3\!/_2\,)(\, y + {}^1\!/_2\,) + {}^5\!/_2 \qquad\qquad (5)$$

The situation is quite different. While A still controls the first factor and B controls the second factor, neither one can make their factor equal to zero. B notices that A's factor, $(\, x - {}^3\!/_2\,)$, is always negative no matter what probability A picks. Therefore, B will make his factor, $(\, y + {}^1\!/_2\,)$, as large as possible to take advantage of the negative product. B will choose $y = 1$. Of course, A wishes to make his negative factor as small as possible. The best he can do is to pick $x = 1$. Thus, the optimal strategies are $(\, 1, 0\,)$ and both players will always play heads.   ●

DEFINITION 8.   A game is called *strictly determined* if the optimal strategies of both players are pure.

Example 3 had optimal pure strategies and is a strictly determined game. From equation $(\, 5\,)$ we see that the value of the game is $z = 1$ for both players. Player A will win at least 1¢ on each play of the game. A might win more than 1¢ if B were to use any strategy other than $(\, 1, 0\,)$. If B uses $(\, 1, 0\,)$ then he is guaranteed to pay out no more than 1¢ to A. The saddle surface.

$$z = 2xy + x - 3y + 1$$

has a saddle point at $(\, {}^3\!/_2, -{}^1\!/_2, {}^5\!/_2\,)$, but in this case its coordinates are outside of the range of probabilities.

The same solution can be reached directly from the pay-off matrix by the following reasoning. The poorest A can do in the first row is to win 1¢ while the worst he can do in the second row is to lose 2¢. He will pick the first row to get the maximum of his minimum possible gains. Player B looks at columns. The most B can lose in the first column is 1¢ while his maximum loss in the second column is 2¢. He will pick the first column to get the minimum of his maximum possible losses. One cent is both the maximum of A's minimums and the minimum of B's maximums. This common value is known as a *minimax* value. A minimax is the common value that both players will pay or receive under optimal pure strategies. Thus, if a matrix has a minimax value, it is the value of the game and the game is strictly determined. While most matrices do not have a minimax entry, we will show that the value of all games is the same for both A and B. That is, the maximum that A will win on the average per play is the same as the minimum *B* will lose on the average per play when each uses his best strategy. This result is the Minimax Theorem proved by John von Neumann.

PROBLEMS

For the following matrix games, find the game value and the optimal strategies by first finding the saddle surface $z = f(x,y)$. Also state whether the game is fair. In case the game is strictly determined, give the location of the minimax entry.

6–1. (a) $\begin{bmatrix} 5 & -4 \\ -3 & 2 \end{bmatrix}$

(b) $\begin{bmatrix} -4 & 3 \\ 6 & 0 \end{bmatrix}$

6–2. (a) $\begin{bmatrix} 0 & 2 \\ -2 & 0 \end{bmatrix}$

(b) $\begin{bmatrix} -5 & 4 \\ 3 & -1 \end{bmatrix}$

6–3. (a) $\begin{bmatrix} 1 & 0 \\ 2 & -3 \end{bmatrix}$

(b) $\begin{bmatrix} 3 & -2 \\ -2 & 3 \end{bmatrix}$

6–4. (a) $\begin{bmatrix} -5 & 6 \\ 5 & -6 \end{bmatrix}$

(b) $\begin{bmatrix} 1 & 2 \\ 3 & 4 \end{bmatrix}$

6–5. (a) $\begin{bmatrix} -3 & 4 \\ 2 & -3 \end{bmatrix}$

(b) $\begin{bmatrix} 2 & -3 \\ -3 & 4 \end{bmatrix}$

(c) What effect does interchanging the rows have upon the results?

(d) Would the corresponding effect be true for interchanging columns?

6–6. (a) $\begin{bmatrix} a & 0 \\ 0 & d \end{bmatrix}$

(b) $\begin{bmatrix} 0 & b \\ c & 0 \end{bmatrix}$

6–7. Given the pay-off matrix

$$\begin{bmatrix} a & b \\ c & d \end{bmatrix}$$

(a) Find the value of the game, z, and the optimal strategies. Assuming that the game is not strictly determined, these quantities may be expressed in terms of the numbers a, b, c, and d.

(b) How is the value z affected by adding the same constant k to each member of the pay-off matrix?

(c) How is the value z affected by multiplying each member of the pay-off matrix by the same constant k?

(d) Are the optimal strategies affected by these two operations in parts (b) and (c)? The important conclusion here is true in general for any size matrix and will be used in Section 6.6.

6.5   A GRAPHICAL SOLUTION OF $m \times 2$ MATRIX GAMES

In the special case of $m \times 2$ matrix games, A's optimal strategy may be found by graphing his strategy polygon. A similar dual analysis will find B's optimal strategy in the case of $2 \times n$ matrix games. Consider the following example.

Example 4.

$$
A \qquad
\begin{array}{c} B \\
\begin{bmatrix}
3 & 1 \\
1 & -2 \\
2 & 6 \\
-4 & 2
\end{bmatrix}
\end{array}
$$

Let A's strategy be (t_1, t_2, t_3, t_4) where $t_i \geq 0$ for $i = 1,2,3,4$ and $t_1 + t_2 + t_3 + t_4 = 1$. First consider the pay-off to A in the extreme cases of B's two possible pure strategies. Let x equal A's pay-off against B's pure strategy (1, 0). Then

$$x = 3t_1 + 1t_2 + 2t_3 - 4t_4. \tag{1}$$

Let y equal A's pay-off against B's pure strategy (0, 1). Then

$$y = 1t_1 - 2t_2 + 6t_3 + 2t_4 \tag{2}$$

If B uses a mixed strategy ($s, 1 - s$) then his pay-off to A, called z, will be the convex combination of pay-offs x and y,

$$z = sx + (1 - s)y.$$

Pay-off z is always between x and y. For example if $x \leq y$, then

$$x = sx + (1 - s)x \leq z \leq sy + (1 - s)y = y.$$

The inequalities are reversed in case $x \geq y$. This proves the following theorem.

THEOREM 6-4.   *B's minimum pay-off to A is the smaller of x or y.*

We wish to graph the set of points (x, y) representing all possible pay-offs to player A under B's pure strategies. The vertices of this set are found by substituting the four possible pure strategies of A into equations (1) and (2). These points correspond respectively to the rows of the pay-off matrix. The entire set, $\{ (x, y) \}$, satisfying the restrictions on t_i, is the smallest convex set containing these four pure strategy points of A. It is the *convex hull* of the pure strategy points. The set may be shown graphically by drawing the smallest convex polygon, R, containing the pure strategy points. The resulting polygon, known as A's *strategy polygon*, is shown in Figure 6–2.

By theorem 6–4, for any point of A's strategy polygon, B will minimize his pay-off to A by picking the smaller of its coordinates x or y. The best choice for A is to pick a strategy point (x, y) whose smaller coordinate is a maximum over R. This may be done by proceeding out from the origin along the line $x = y$ until it intersects the polygon boundary at a point with the largest abscissa x. In our example, that boundary has a negative slope and the desired point is the intersection z shown. Points above line $x = y$ in the polygon have a smaller abscissa and points below $x = y$ in the polygon have a smaller ordinate than z. If the boundary in question has a positive slope, then we may continue out along this boundary until we reach the vertex with the largest abscissa. In that case the game would be strictly determined.

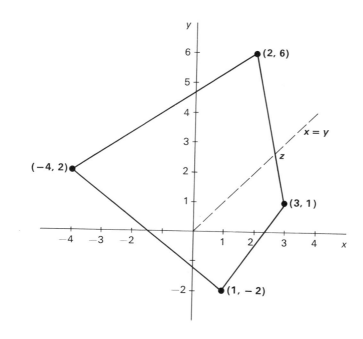

Figure 6–2

In our case, z represents the optimum pay-off, and either one of its coordinates gives the value of the game for A. The parametric equations of the boundary from (3, 1) to (2, 6) are

$$x = 3t_1 + 2(1 - t_1)$$
$$y = 1t_1 + 6(1 - t_1)$$
$$t_2 = 0, \; t_4 = 0, \; t_3 = 1 - t_1.$$

Probabilities t_2 and t_4 are 0 since the boundary line of R through z is a convex combination of vertices (3, 1) and (2, 6). In the parametric equations, note that probability $t_1 = 1$ corresponds to vertex (3, 1) and $t_1 = 0$ or $t_3 = 1$ corresponds to vertex (2, 6).

Setting x equal to y and solving for t_1, we find $t_1 = \frac{2}{3}$, $t_3 = (1 - t_1) = \frac{1}{3}$. Thus A's optimal strategy is ($\frac{2}{3}$, 0, $\frac{1}{3}$, 0). The game value for A is $x = y = \frac{8}{3}$. By the Minimax Theorem $\frac{8}{3}$ is also the game value for player B. ●

Since the game value will always occur at a boundary point of A's strategy polygon, we have the following theorem.

THEOREM 6-5. *In $m \times 2$ matrix games player A needs at most a convex combination of two of his m pure strategies. Similarly, in $2 \times n$ matrix games, B needs at most a convex combination of two of his n pure strategies.*

An alternate graph to the strategy polygon is the line graph in the following example.

Example 5. Consider the following $2 \times n$ game.

$$
\begin{array}{c}
 & B \\
A & \begin{array}{|rrrr|}
\hline
5 & -2 & 2 & -1 \\
-2 & 4 & -1 & 1 \\
\hline
\end{array}
\end{array}
$$

Let A's strategy be ($1 - t, t$). Let x_j, $j = 1, 2, 3, 4$, be the pay-offs to A under B's four possible pure strategies. Then A wishes to maximize the minimum of the linear functions

$$x_j = Y_{1j}(1 - t) + Y_{2j}t$$

for $j = 1, 2, 3, 4$ where Y_{ij} are the entries in the pay-off matrix. These linear functions of t are respectively

$$x_1 = \;\;\; 5(1 - t) - 2t$$
$$x_2 = -2(1 - t) + 4t$$
$$x_3 = \;\;\; 2(1 - t) - t$$
$$x_4 = \;\; -(1 - t) + t.$$

The four lines are graphed in Figure 6–3; note that each line passes through (0, Y_{1j}) and (1, Y_{2j}). The minimum function of these lines is the sequence of line segments in bold in Figure 6–3. The maximum of this minimum function occurs at the intersection of lines x_3 and x_4. Setting $x_3 = x_4$ gives

$$2(1 - t) - t = -(1 - t) + t$$
$$2 - 3t = -1 + 2t$$
$$t = \tfrac{3}{5}.$$

Thus A's optimal strategy is ($\tfrac{2}{5}$, $\tfrac{3}{5}$) and the value of the game is $x_3 = x_4 = \tfrac{1}{5}$.

B's optimal strategy may be found by finding the ratio in which height $\tfrac{1}{5}$ divides the line segment PQ from x_4 to x_3 at $t = 0$. This ratio is

$$(\tfrac{6}{5})/(\tfrac{9}{5}) = \tfrac{2}{3}.$$

The strategy for B corresponding to the ratio of 2 to 3 is (0, 0, $\tfrac{2}{5}$, $\tfrac{3}{5}$). The same strategy could be found by using the ratio in which height $\tfrac{1}{5}$ divides the line segment RS from x_3 to x_4 at $t = 1$. The ratio is

$$(\tfrac{6}{5})/(\tfrac{4}{5}) = \tfrac{3}{2}.$$

The ratio is reversed since the order of the corresponding lines is reversed at $t = 1$. ●

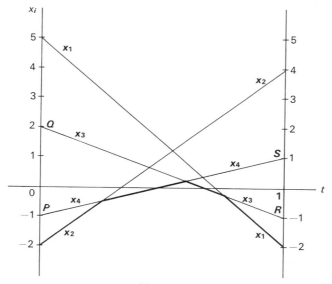

Figure 6–3

In the case of $m \times 2$ matrix games, the corresponding lines would be drawn for A's m pure strategies. Then the game value would be the minimum height of the maximum of these linear functions.

PROBLEMS

For the following matrix games find the game value and player A's optimal strategy from A's strategy polygon.

6–8. $\begin{bmatrix} 4 & 1 \\ -1 & 3 \end{bmatrix}$

6–9. (a) $\begin{bmatrix} -1 & 1 \\ -3 & 7 \\ 3 & -2 \\ 1 & 6 \end{bmatrix}$

(b) $\begin{bmatrix} -1 & 1 \\ -3 & 7 \\ 2 & 5 \\ 1 & -1 \end{bmatrix}$

6–10. (a) $\begin{bmatrix} 2 & 4 \\ 5 & 1 \\ 3 & 6 \end{bmatrix}$

(b) $\begin{bmatrix} -5 & 1 \\ 3 & -2 \\ 4 & 7 \end{bmatrix}$

6–11. (a) $\begin{bmatrix} -8 & 5 \\ 2 & -4 \\ 6 & -1 \end{bmatrix}$

(b) $\begin{bmatrix} 4 & 1 \\ 3 & -2 \\ 2 & -4 \end{bmatrix}$

6–12. $\begin{bmatrix} 1 & 6 \\ -2 & 3 \\ -1 & 0 \\ 2 & -4 \\ 7 & 3 \end{bmatrix}$

For the next set of matrix games, find the game value and player B's optimal strategy from B's strategy polygon. In this case choose the point of B's polygon for which the larger coordinate is minimized.

6–13. $\begin{bmatrix} 4 & 1 \\ -1 & 3 \end{bmatrix}$

6–14. (a) $\begin{bmatrix} 0 & 3 & 4 \\ 3 & 2 & 0 \end{bmatrix}$

(b) $\begin{bmatrix} 2 & -1 & 5 \\ 5 & -2 & -1 \end{bmatrix}$

6–15. (a) $\begin{bmatrix} -2 & 3 & -1 \\ 4 & 3 & 1 \end{bmatrix}$

(b) $\begin{bmatrix} 0 & -4 & 4 \\ 3 & 1 & -3 \end{bmatrix}$

6–16. $\begin{bmatrix} -8 & 1 & 4 & 6 \\ 1 & 7 & -3 & 2 \end{bmatrix}$

6–17. $\begin{bmatrix} -4 & 2 & 2 & -2 & 5 \\ 1 & 5 & -1 & -3 & -2 \end{bmatrix}$

6–18. Solve problem 5–16 by drawing the appropriate line graph.

6–19. Solve problem 5–17 by drawing its line graph.

6–20. Solve problem 5–12 by drawing its line graph.

6.6 LINEAR PROGRAMMING FOR MATRIX GAMES AND THE DEFINITION OF DUALITY

In this section we will develop a linear program to solve the general $m \times n$ matrix game. In the process we will discover the underlying principle of duality. The Fundamental Duality Theorem will then imply von Neumann's Minimax Theorem as a corollary.

Example 6.   Let us solve the following matrix game.

$$A \begin{bmatrix} & & B & \\ 1 & -2 & 1 \\ -1 & 3 & -2 \\ -1 & -2 & 3 \end{bmatrix}$$

There are two problems. A's problem is to find his optimal strategy and his maximum pay-off under optimal strategies. B's problem is to determine his optimal strategy and his minimum pay-off to A under optimal strategies. Let (x_1, x_2, x_3) be B's strategy and let (x_4, x_5, x_6) be A's strategy where $x_i \geq 0$, $i = 1, 2, \ldots, 6$, and

$$x_1 + x_2 + x_3 = 1 \tag{1}$$

$$x_4 + x_5 + x_6 = 1. \tag{2}$$

Because it will be easier to work with a positive pay-off, we add 2 to each entry in the pay-off matrix. This will remove all negative entries and ensure that the pay-off to A is a positive number. Recall, from the problems of section 6.4, that the addition of a constant to the pay-off matrix does not change the optimal strategies. However, the game value will be 2 more than the value of the original game. Thus we will solve the following game and then subtract 2 from its game value.

$$
\begin{array}{c}
 & B \\
A & \begin{bmatrix} 3 & 0 & 3 \\ 1 & 5 & 0 \\ 1 & 0 & 5 \end{bmatrix}
\end{array}
$$

We attack B's problem first. Let L_1, L_2, L_3 be respectively B's pay-offs to A under A's three pure strategies, $(1, 0, 0)$, $(0, 1, 0)$, and $(0, 0, 1)$. Then

$$
\begin{aligned}
L_1 &= 3x_1 + 0x_2 + 3x_3 \\
L_2 &= 1x_1 + 5x_2 + 0x_3 \\
L_3 &= 1x_1 + 0x_2 + 5x_3.
\end{aligned}
\tag{3}
$$

Let L be the largest of the three pay-offs, L_1, L_2, and L_3. Since A will choose L, B wishes to minimize L. By replacing the L_i in equations (3) with L, we have type I linear constraints

$$
\begin{aligned}
3x_1 \qquad + 3x_3 &\le L \\
1x_1 + 5x_2 \qquad &\le L \\
1x_1 \qquad + 5x_3 &\le L.
\end{aligned}
\tag{4}
$$

Consider each of the constraints in (4) divided by L. The pay-offs to A were forced to be positive so that L would be greater than zero. Thus the sense of the inequalities is preserved. Now rename the variables as follows:

$$
x_1' = x_1/L, \quad x_2' = x_2/L, \quad x_3' = x_3/L.
$$

Also divide equation (1) through by L and then make the given change of variables. For B to minimize L, he will maximize $1/L$. B's problem may now be stated as a linear program.

Find nonnegative numbers x_1', x_2', x_3', subject to the constraints

$$
\begin{aligned}
3x_1' \qquad + 3x_3' &\le 1 \\
1x_1' + 5x_2' \qquad &\le 1 \\
1x_1' \qquad + 5x_3' &\le 1,
\end{aligned}
\tag{5}
$$

such that the number

$$
x_1' + x_2' + x_3' = 1/L = M
$$

is a maximum.

Let the three slack variables be x_4', x_5', and x_6'. Then B's problem leads directly to the considered tableau

	1	2	3	
4	3	0	3	1
5	1	5	0	1
6	1	0	5	1
−1	−1	−1	0

Note that the tableau is made up of the pay-off matrix, with a column on 1s on the right, and a row of minus 1s at the bottom along with zero as the initial value of M. Three pivots bring us to the optimal tableau. The first pivot is on position (1, 1) giving

	4	2	3	
1	$1/3$	0	1	$1/3$
5	$-1/3$	5	−1	$2/3$
6	$-1/3$	0	4	$2/3$
	$1/3$	−1	0	$1/3$

Next, pivot on (2, 2).

	4	5	3	
1	$1/3$	0	1	$1/3$
2	$-1/15$	$1/5$	$-1/5$	$2/15$
6	$-1/3$	0	4	$2/3$
	$4/15$	$1/5$	$-1/5$	$7/15$

Finally, pivot on (3, 3) for the following optimal tableau.

	4	5	6	
1	$5/12$	0	$-1/4$	$1/6$
2	$-1/12$	$1/5$	$1/20$	$1/6$
3	$-1/12$	0	$1/4$	$1/6$
	$1/4$	$1/5$	$1/20$	$1/2$

From the optimal tableau $M = 1/2$ and $L = 1/M = 2$. From our transformation equations

$$x_1 = Lx_1' = 2(1/6) = 1/3$$
$$x_2 = Lx_2' = 2(1/6) = 1/3$$
$$x_3 = Lx_3' = 2(1/6) = 1/3.$$

Thus B's optimal strategy is ($1/3$, $1/3$, $1/3$). Remembering that 2 was added to the original pay-off matrix, we find the game value from B's viewpoint to be $L - 2 = 0$. Zero is the minimum B may expect to pay A on the average per play of the game. Thus the game is fair to B.

Next let us solve A's problem. Let S_1, S_2, S_3 be respectively A's pay-offs under B's three pure strategies (1, 0, 0), (0, 1, 0), and (0, 0, 1). Then

$$S_1 = 3x_4 + 1x_5 + 1x_6$$
$$S_2 = 0x_4 + 5x_5 + 0x_6 \qquad\qquad (6)$$
$$S_3 = 3x_4 + 0x_5 + 5x_6.$$

Let S be the smallest of the three pay-offs, S_1, S_2, and S_3. Since B will choose S, A wishes to maximize S. By replacing the S_i in equations (6) with S, we have type II inequalities.

$$3x_4 + 1x_5 + 1x_6 \geq S$$
$$5x_5 \qquad\quad \geq S \qquad\qquad (7)$$
$$3x_4 \qquad\; + 5x_6 \geq S.$$

Consider each of the constraints in (7) divided by S. Once again the sense of the inequalities is preserved because all pay-offs were forced to be positive and $S > 0$. Rename the variables as follows,

$$x_4' = x_4/S, \qquad x_5' = x_5/S, \qquad x_6' = x_6/S.$$

Also divide equation (2) through by S and make the change of variables. For A to maximize S, he will minimize $1/S$. A's problem may now be stated as a linear program.

Find nonnegative numbers x_4', x_5', x_6', subject to the constraints

$$3x_4' + 1x_5' + 1x_6' \geq 1$$
$$5x_5' \qquad\quad \geq 1 \qquad\qquad (8)$$
$$3x_4' \qquad\; + 5x_6' \geq 1,$$

and such that the number

$$x_4' + x_5' + x_6' = 1/S = m$$

is a minimum.

Let the three slack variables be x_1', x_2', x_3', and let the three artificial variables be x_{-7}, x_{-8}, x_{-9}. Change the minimizing problem to a maximizing problem by setting $m = -M$. Let the artificial maximum be

$$\bar{M} = M - N(x_{-7} + x_{-8} + x_{-9})$$

or

$$\bar{M} + x_4' + x_5' + x_6' + N(x_{-7} + x_{-8} + x_{-9}) = 0.$$

N is an arbitrarily large positive number so that the artificial variables will leave the basis and then $\bar{M}$ will equal M. This leads to the following initial condensed tableau for A.

	4	5	6	1	2	3	
-7	3	1	1	-1	0	0	1
-8	0	5	0	0	-1	0	1
-9	3	0	5	0	0	-1	1
$\bar{M}$	1	1	1	0	0	0	0
	-6	-6	-6	1	1	1	-3

We will break the ties by pivoting on (1, 1) with the following result.

	-7	5	6	1	2	3	
4	.	$1/3$	$1/3$	$-1/3$	0	0	$1/3$
-8	.	5	0	0	-1	0	1
-9	.	-1	4	1	0	-1	0
$\bar{M}$	.	$2/3$	$2/3$	$1/3$	0	0	$-1/3$
	.	-4	-4	-1	1	1	-1

Next, pivot on (2, 2) and drop the first column.

	-8	6	1	2	3	
4	.	$1/3$	$-1/3$	$1/15$	0	$4/15$
5	.	0	0	$-1/5$	0	$1/5$
-9	.	4	1	$-1/5$	-1	$1/5$
$\bar{M}$	.	$2/3$	$1/3$	$2/15$	0	$-7/15$
	.	-4	-1	$1/5$	1	$-1/5$

Again drop the first column and then pivot on the first entry of the third row. This pivot corresponds to position (3, 3) of the original matrix and produces the final tableau.

		-9	1	2	3	
		$-5/12$	$1/12$	$1/12$		$1/4$
4	.	$-5/12$	$1/12$	$1/12$	$1/4$	
5	.	0	$-1/5$	0	$1/5$	
6	.	$1/4$	$-1/20$	$-1/4$	$1/20$	
$\overline{M} = M$	.	$1/6$	$1/6$	$1/6$	$-1/2$	
	.	0	0	0	0	

The artificial variables have now been romoved from the basis and may be ignored along with the bottom row. In particular $\overline{M} = M$. Rewrite the optimal tableau accordingly.

	1	2	3	
4	$-5/12$	$1/12$	$1/12$	$1/4$
5	0	$-1/5$	0	$1/5$
6	$1/4$	$-1/20$	$-1/4$	$1/20$
M	$1/6$	$1/6$	$1/6$	$-1/2$

The solution to A's problem may now be read off. $M = -1/2$ so the minimum, $m = -M = 1/2$. Then $S = 1/m = 2$.

Transforming back to the original variables, we have

$x_4 = Sx_4' = 2(1/4) = 1/2$

$x_5 = Sx_5' = 2(1/5) = 2/5$

$x_6 = Sx_6' = 2(1/20) = (1/10)$.

Thus A's optimal strategy is ($1/2$, $2/5$, $1/10$). As before we must subtract 2 from the game value to get back to the original game. A's game value is $S - 2 = 0$. Zero is the maximum A may expect to win on the average per play of the game. The game is fair to both players. As is predictable by the Minimax Theorem, A's maximum gain is the same as B's minimum loss under optimal strategies. ●

Far more important than the solution to the game is the comparison between A's final tableau and B's final tableau. Carefully compare the two.

	4	5	6	
1	$5/12$	0	$-1/4$	$1/6$
2	$-1/12$	$1/5$	$1/20$	$1/6$
3	$-1/12$	0	$1/4$	$1/6$
	$1/4$	$1/5$	$1/20$	$1/2$

	1	2	3	
4	$-5/12$	$1/12$	$1/12$	$1/4$
5	0	$-1/5$	0	$1/5$
6	$1/4$	$-1/20$	$-1/4$	$1/20$
	$1/6$	$1/6$	$1/6$	$-1/2$

B's optimal tableau A's optimal tableau

It was not by accident that we labeled the slack variables of one problem the same as the main variables of the other problem. The fact that the slack and main variables are interchanged is a part of the duality picture. It should be noticed that the values of all six primed variables can be read off, directly opposite their subscripts, from either one of the two optimal tableaux. In other words, the solution to either of the problems automatically gives the solution to both. The matrix in the upper left hand corner of one tableau is the *negative transpose* of the other. That is, any row of one matrix is the negative of the corresponding column of the other. Finally the optimum value in the lower right corner of one tableau is the negative of the optimum value in the other tableau. It is time to define dual tableaux.

DEFINITION 9. The *dual* of an $m \times n$ *tableau* is an $n \times m$ tableau satisfying the following four conditions:

1. The basic and the nonbasic variables are interchanged. This interchanges the subscripts $L(I)$ with $K(J)$.
2. The matrix in the upper left-hand corner is the negative transpose of the corresponding matrix.
3. The bottom row and right-hand column are interchanged.
4. The value in the lower right-hand corner is the negative of its corresponding value.

The definition of dual tableaux is clearly symmetric and involutory. *Symmetric* means that if T_1 is dual to T_2 then T_2 is dual to T_1. *Involutory* means that the dual of the dual is the original tableau.

Recall that the process of pivoting was derived from Gauss-Jordan elimination. Thus, we are able to preserve equivalent systems of equations from tableau to tableau. Therefore, the set of linear equations represented in our final tableau is equivalent to the original set of equations in the initial tableau. Now if two problems have their final tableaux the dual of one another, we might suspect that the original tableaux were dual. To compare the initial tableau of player A's problem with that of B, we first avoid artificial variables by multiplying the constraints (8) by minus one. This leads to the following statement of A's problem.

Example 7.   Find nonnegative numbers x_4', x_5', x_6', subject to the constraints

$$-3x_4' - 1x_5' - 1x_6' \leq -1$$
$$- 5x_5' \leq -1 \qquad\qquad (9)$$
$$-3x_4' - 5x_6' \leq -1,$$

and such that $x_4' + x_5' + x_6' = m = -\overline{M}$, where m is a minimum and $\overline{M}$ is a maximum. Using constraints (5) and (9) the two initial tableaux appear as

	1	2	3	
4	3	0	3	1
5	1	5	0	1
6	1	0	5	1
M	-1	-1	-1	0

B's initial tableau

	4	5	6	
1	-3	-1	-1	-1
2	0	-5	0	-1
3	-3	0	-5	-1
$\overline{M}$	1	1	1	0

A's initial tableau

Indeed the two initial tableaux are dual and we will define the two problems to be dual linear programs.   ●

DEFINITION 10.   Two linear programs are *dual* provided: the slack variables of one are the main variables of the other; type I and type II inequalities are interchanged; the entire augmented matrix, including the objective function, of one is the transpose of the other; and finally the objective maximization is interchanged with minimization.

As an example of definition 10, the following two programs are dual. Find $x_1 \geq 0$, $x_2 \geq 0$, $x_3 \geq 0$ satisfying

$$a_{11}x_1 + a_{12}x_2 + a_{13}x_3 \leq b_1$$
$$a_{21}x_1 + a_{22}x_2 + a_{23}x_3 \leq b_2,$$

and

$$c_1x_1 + c_2x_2 + c_3x_3 = M, \qquad \text{a maximum.}$$

Find $x_4 \geq 0$, $x_5 \geq 0$ satisfying

$$a_{11}x_4 + a_{21}x_5 \geq c_1$$

$$a_{12}x_4 + a_{22}x_5 \geq c_2$$

$$a_{13}x_4 + a_{23}x_5 \geq c_3,$$

and

$$b_1x_4 + b_2x_5 = m, \qquad \text{a minimum.}$$

If the initial tableaux of the previous problems are written without artificial variables, they will satisfy the duality of definition 9. Assuming that all of the constraints of the maximization problem are type I inequalities, the two definitions 9 and 10 are equivalent. Duality for problems with a mixture of constraints is discussed in Chapter 8.

6.7 THE DUAL SIMPLEX METHOD

In trying to solve player A's problem without artificial variables, we run into a difficulty. A's initial tableau is already optimal but not feasible. The initial value of the basic variables x_1', x_2', x_3' is -1. We wish to transform this tableau by a method, known as the *Dual Simplex Method*, that will gain feasibility while preserving the optimality. The Dual Simplex Method utilizes pivoting, but the pivot will be chosen by the following algorithm.

THE DUAL SIMPLEX ALGORITHM

1. Starting with a condensed tableau that is optimal but not feasible, choose the pivotal row by picking the most negative value in the last column from the first $m - 1$ rows.

2. Using only the negative entries in that pivotal row, from the first $n - 1$ columns, compute the θ ratios. They are the ratios of the entries in the last row to the corresponding entries in the pivotal row. All of these ratios are negative.

3. Determine the pivotal column by picking the algebraically largest ratio of step (2). This will be the ratio whose absolute value is the smallest. In case of ties, pick any one of the tying columns.

4. Carry out a pivoting iteration by the condensed tableau algorithm of Chapter 4 using the pivot chosen by steps 1 and 3.

5. Repeat the first four steps until the tableau is feasible or until no new pivot can be found.

To distinguish a program from its dual, we call the maximization problem the *primal program* and call its dual the *dual program*. The dual program is then a minimization problem. In the case of our matrix game with players A and

B, the primal program is B's problem since his initial tableau corresponds to a maximization, while A's problem is the dual program since his initial tableau corresponds to a minimization. The pivot chosen by the Dual Simplex Algorithm in the dual program is the *dual* of the corresponding pivot chosen by the Simplex Algorithm in the primal program. This means that if the pivot is at (i, j) in the primal tableau then it will occur at (j, i) in the dual tableau. If the initial tableau of player B is pivoted by the Simplex Method, and if the dual initial tableau of player A is pivoted by the Dual Simplex Method, then each pair of tableaux will be dual through the final pair. This important idea should be checked out on the tableaux of A and B in the game problem of the last section. We will prove in the next chapter that dual pivots of a pair of dual tableaux lead to a pair of dual tableaux. This is the heart of the Fundamental Duality Theorem.

PROBLEMS

Solve the following matrix games by pivoting the condensed tableaux. State the optimal strategies of both players and the value of the game.

6–21. (a) $\begin{bmatrix} 3 & 5 \\ 2 & 1 \\ 5 & 2 \end{bmatrix}$ (b) $\begin{bmatrix} 5 & -2 \\ 4 & 6 \\ -3 & 4 \end{bmatrix}$

6–22. (a) $\begin{bmatrix} 0 & 3 & 5 \\ 1 & 0 & 3 \end{bmatrix}$ (b) $\begin{bmatrix} -1 & 5 & 4 \\ 3 & 6 & -1 \end{bmatrix}$

6–23. (a) $\begin{bmatrix} 1 & 2 \\ 3 & 4 \\ 5 & 3 \end{bmatrix}$ (b) $\begin{bmatrix} 2 & -2 & -4 \\ -3 & 2 & 3 \end{bmatrix}$

6–24. (a) $\begin{bmatrix} 2 & 1 \\ -2 & 6 \\ 3 & 5 \end{bmatrix}$ (b) $\begin{bmatrix} 3 & 4 & 6 \\ 1 & 5 & 4 \end{bmatrix}$

6–25. (a) $\begin{bmatrix} 5 & 3 & 2 \\ 3 & 4 & 1 \end{bmatrix}$ (b) $\begin{bmatrix} 4 & 2 \\ 1 & 3 \\ 3 & 4 \end{bmatrix}$

(c) Check the solutions to (a) and (b) by solving them graphically.

6–26. Solve the following matrix game for optimal strategies by setting up A's problem and then pivoting according to the Dual Simplex Algorithm.

$$\begin{bmatrix} 1 & -2 & 2 \\ -3 & 4 & -1 \\ 5 & -2 & -1 \end{bmatrix}$$

6-27. Carry out the same directions in problem 6-26 on the following game.

$$\begin{bmatrix} 1 & -1 & 0 \\ 0 & 1 & -1 \\ -1 & 0 & 1 \\ 1 & 0 & -1 \\ -1 & 1 & 1 \end{bmatrix}$$

6-28. Solve the following game for optimal strategies and game value.

$$\begin{bmatrix} 1 & 1 & -1 & 0 & 1 \\ 1 & 0 & 0 & 1 & -1 \\ -1 & 0 & 1 & -1 & 0 \\ 0 & 1 & -1 & 0 & 1 \\ 1 & -1 & 0 & 1 & 1 \end{bmatrix}$$

6-29. Solve the following skew-symmetric[5] matrix game for optimal strategies and game value.

$$\begin{bmatrix} 0 & 1 & -2 & 3 & -4 \\ -1 & 0 & 3 & -4 & 5 \\ 2 & -3 & 0 & 5 & -6 \\ -3 & 4 & -5 & 0 & 7 \\ 4 & -5 & 6 & -7 & 0 \end{bmatrix}$$

Hint: Add 10 to each pay-off entry.

6-30. From problem 6-29 what might you conjecture about the game value and the optimal strategies in the case of a skew-symmetric pay-off matrix? Can you prove your conjecture?

6-31. In a popular childrens' game, *Stone, Scissors, and Paper*, each player independently picks one of the three items. If there is a tie then there is no pay-off. Otherwise stone "breaks" scissors and stone is paid one, paper "covers" stone and paper is paid two, while scissors "cuts" paper and scissors is paid three. Devise a 3 × 3 pay-off matrix for this game. Solve the game for the optimal strategies of both players. Is the game fair?

6-32. Two players independently choose a number from one to three. The pay-off is computed by adding the two numbers chosen. If the sum is even, player *B* pays player *A* that amount, and if the sum is odd, then *A* pays *B* that amount. Set up the pay-off matrix and solve the game for optimal strategies. Is this game fair?

6-33. Two players independently choose a number from one to four. The pay-off is computed by adding the two numbers chosen. If the sum is even, player *B* pays player *A* that amount, and if the sum is odd, then *A* pays *B* that amount. Set up the pay-off matrix and solve the game for optimal strategies. Is this a fair game? Are the optimal strategies unique?

---

[5] A square matrix is skew-symmetric if $Y_{ij} = - Y_{ji}$ for all i and j.

6-34. Two players independently choose a number from one to four. The pay-off is computed by taking the absolute value of the difference of the two numbers chosen. If this absolute value is even, player *B* pays player *A* that amount, and if this absolute value is odd, then *A* pays *B* that amount. Set up the pay-off matrix for this game. Solve the game for its optimal strategies and game value.

6-35. Two players independently choose a number from one to four. If there is a tie between the numbers chosen, then player *B* pays player *A* that tying amount. If the numbers chosen are different then the absolute value of this difference is computed. When this absolute value is even, player *B* pays player *A* that amount, and when this absolute value is odd, *A* pays *B* that amount. Develop the pay-off matrix for this game. Solve the game for its optimal strategies and game value.

CHAPTER 7

Duality Theorems and Revised Simplex Method

7.1 THE FUNDAMENTAL DUALITY THEOREM

In Chapter 6 we found that the solution of a matrix game for the strategies of each player led to a pair of dual linear programs. Furthermore, we noticed that the solution to both programs could be found from the solution to either one of the two dual programs. We will now prove this basic idea in general for any pair of dual linear programs.

THEOREM 7-1. *(Fundamental Duality Theorem) For a pair of dual linear programs, if one solution exists, the solutions to both programs may be found by solving either one of the two programs.*

PROOF. The first job is to show that dual pivots transform dual programs into a new pair of dual programs. Consider the following pair of dual condensed tableaux.

Primal

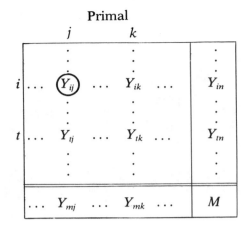

Dual

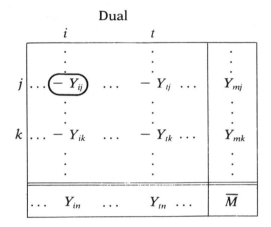

Pivoting the primal tableau at position (i, j) gives

	i	k	
j	$\ldots\quad 1/Y_{ij} \quad\ldots$	$Y_{ik}/Y_{ij} \quad\ldots$	Y_{in}/Y_{ij}
t	$\ldots\quad -Y_{tj}/Y_{ij} \quad\ldots$	$Y_{tk} - Y_{tj}(Y_{ik}/Y_{ij}) \quad\ldots$	$Y_{tn} - Y_{tj}(Y_{in}/Y_{ij})$
	$\ldots\quad -Y_{mj}/Y_{ij} \quad\ldots$	$Y_{mk} - Y_{mj}(Y_{ik}/Y_{ij}) \quad\ldots$	$M - Y_{mj}(Y_{in}/Y_{ij})$

Pivoting the dual tableau at the dual position (j, i) gives

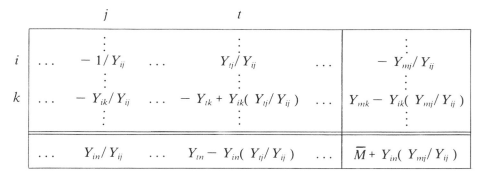

The pivoted tableaux are clearly dual. Notice especially that $M + \overline{M}$ stays constant in the pivoting, that is, the sum of the two objective values is the same after pivoting. Let the transforms of M and $\overline{M}$ under pivoting be $T(\,M\,)$ and $T(\,\overline{M}\,)$ respectively. From the definition of dual tableaux, $M = -\overline{M}$ or $M + \overline{M} = 0$. Then

$$T(\,M\,) + T(\,\overline{M}\,) = M - Y_{in}(\,Y_{in}/Y_{ij}\,) + \overline{M} + Y_{in}(\,Y_{mj}/Y_{ij}\,)$$
$$= M + \overline{M}$$
$$= 0.$$

Thus $T(\,M\,) = -T(\,\overline{M}\,)$. When we have arrived at the feasible optimum for the dual tableau, $T(\,\overline{M}\,)$ is a maximum and therefore $-T(\,\overline{M}\,)$ is a minimum. Remember that in the dual program we replace the objective function to be minimized with the corresponding function $\overline{M}$ to be maximized, in order to use the same pivoting algorithm in both cases.

The next job is to show that when one tableau arrives at its feasible optimum, the dual tableau must also be at its feasible optimum. Suppose the pivoted primal tableau here is optimum and its objective maximum is $T(\,M\,)$. Then the first $n-1$ entries in its last row are positive, and since each pivot preserves feasibility the first $m-1$ entries in its right-hand column are similarly positive. Now the corresponding dual tableau must be feasible because its right-hand column contains respectively the $n-1$ positive numbers of the last row in the primal; and the dual tableau remains optimal since its last row contains respectively the $m-1$ positive numbers of the right-hand column in the primal. Thus the pivoting of the dual tableau is complete and $-T(\,\overline{M}\,)$ is its objective minimum. But we have shown $T(\,M\,) = -T(\,\overline{M}\,) = m$, so that the maximum of the primal is the minimum of the dual. Finally this interchange of bottom row with right-hand column shows that both sets of numbers may be found from either one of the two optimized tableaux. ∎

7.2   THE MINIMAX THEOREM

THEOREM 7-2.   *(The Minimax Theorem) The maximum value of the objective function in the solution to the primal program is the same as the minimum value of the objective function in the solution to the dual program.*

In the context of game theory the Minimax Theorem says that the maximum of A's minimum expectation is the same as the minimum of B's maximum expectation. In other words A's game value is the same as B's game value. Using only elementary techniques we have proved the von Neumann Minimax Theorem not only for zero-sum matrix games but also for general linear programs. We have in fact a choice of dual procedures for solving a linear program. We may solve the primal program by pivoting according to the Simplex Algorithm, or we may solve the dual program by pivoting according to the Dual Simplex Algorithm. It is wise to pick the method that leads to the simpler of the two solutions. Although the two methods are equivalent to each other, one may be easier to set up than the other.

The symmetric dualization developed so far is appropriate for problems involving constraints that are all of type I or are all of type II. The situation for equality constraints and a mixture of types is more complicated. An equality is equivalent to a pair of inequalities, one of each type. Therefore, an equality constraint may be replaced with two inequality constraints, one of type I and the other of type II. This necessarily leads to a system of mixed constraints. Mixed systems will be discussed in the next chapter.

7.3   THE REVISED SIMPLEX METHOD

The pivoting computations of the Simplex Algorithm can be carried out by matrix multiplication. The Revised Simplex Method does the job with matrix algebra and carries along only the essential information at each iteration. All that is needed to move from one iteration to the next is the pivotal column, the right-hand column, the objective row, and a representation of the inverse of a basic matrix.

Consider the extended tableau of section 4.2 for a maximization problem with type I inequalities. The slack variables are in the basis of the initial tableau. Place the objective value, M, into the basis by adding another column unit vector with a 1 in the objective row. Let $X_1, X_2, \ldots, X_{n-1}$ be the set of problem variables including slack variables. Then $X_n = M$ may represent the objective value, and be the label of the new column. Thus, the initial tableau of section 4.2 is

X_1	X_2	X_3	X_4	X_5	
1	3	1	0	0	7
2	1	0	1	0	4
-1	-2	0	0	1	0

This tableau is the same as the augmented matrix of the corresponding system of equations.

Let m equal the total number of rows, that is, the number of constraints plus one for the objective row. In matrix notation the tableau may be written

$$\mathbf{AX} = \mathbf{C} \tag{1}$$

where $\mathbf{A}$ is the m by n coefficient matrix, $\mathbf{X}$ is the n by 1 column vector of unknowns, and $\mathbf{C}$ is the m by 1 column vector of constants on the right-hand side of each equation. Provided the slack variables are in the basis, matrix $\mathbf{A}$ contains an identity submatrix that is m by m and corresponds to the basic columns of $\mathbf{A}$.

If artificial variables are necessary, then some changes must be made to accommodate the additional variables. Let $\mathbf{A}_q$ represent the qth column of $\mathbf{A}$ and partition $\mathbf{A}$ into two parts where $\mathbf{B}_j$ represents the columns of $\mathbf{A}$ that are in the basis. Then columnwise

$$\mathbf{A} = \mathbf{A}_1 \mathbf{A}_2 \ldots \mathbf{A}_{n-m} \mathbf{B}_1 \mathbf{B}_2 \ldots \mathbf{B}_m.$$

In equation (1) the nonbasic variables are set equal to zero. If $\mathbf{X}_B$ is the column vector of basic variables, then (1) reduces to

$$\mathbf{B}\,\mathbf{X}_B = \mathbf{C} \text{ or}$$

$$\mathbf{X}_B = \mathbf{B}^{-1}\,\mathbf{C}. \tag{2}$$

Since $\mathbf{B}$ initially is the identity matrix, $\mathbf{B}^{-1}$ also is the identity and the first basic feasible solution is $\mathbf{X}_B = \mathbf{C}$, where $X_n = C_m = M$, the current objective value.

Equation (2) is true for any set of basic variables that may appear in a subsequent tableau where $\mathbf{B}$ contains the columns of those basic variables in the initial tableau. This follows because the basic variables in any tableau represent a basic solution to the original system, and therefore their columns of coefficients must be linearly independent, that is, have rank m. Thus $\mathbf{B}$ is nonsingular and has an inverse.

The Revised Simplex Method offers a convenient way to compute $\mathbf{B}^{-1}$, called the product form of the inverse. In the chapter on linear algebra, it was shown that the inverse of a nonsingular matrix $\mathbf{B}$ could be found by pivoting. Each part of a pivoting operation on $\mathbf{B}$ could be carried out by multiplying $\mathbf{B}$ on the left by an elementary matrix. At the end of m pivots, if there were k such elementary operations, then the value of the inverse is

$$\mathbf{B}^{-1} = \mathbf{E}_k \mathbf{E}_{k-1} \ldots \mathbf{E}_2 \mathbf{E}_1 \mathbf{I}$$

where $\mathbf{I}$ is the m by m identity.

Start with $\mathbf{B}^{-1} = \mathbf{I}$ in the initial tableau. Let $\mathbf{T}$ be the matrix that combines the elementary operations in a pivot from the initial tableau to the second tableau. Then the inverse of the new base in tableau two is

$$\overline{\mathbf{B}}^{-1} = \mathbf{T}\,\mathbf{B}^{-1} = \mathbf{T}.$$

Similarly, letting $\mathbf{T}$ be the elementary matrix corresponding to any pivoting iteration, the new basis inverse may be found from the previous $\mathbf{B}^{-1}$ by

$$\overline{\mathbf{B}}^{-1} = \mathbf{T}\,\mathbf{B}^{-1}. \tag{3}$$

A method to find the elementary matrix $\mathbf{T}$ at each iteration is necessary.

First let us prove that any column in a subsequent simplex tableau may be found by multiplying the corresponding column in the initial tableau by the inverse of the current basic matrix. That is,

$$\overline{\mathbf{A}}_q = \mathbf{B}^{-1}\mathbf{A}_q \text{ for } q = 1, 2, \ldots, n+1. \tag{4}$$

Any column of $\mathbf{A}$ in the initial tableau can be written as a linear combination of the columns of $\mathbf{B}$, since the columns of $\mathbf{B}$ are linearly independent and span the space. The $\mathbf{B}$ columns are said to form a basis for the m dimensional space.

$$\mathbf{A}_q = d_1\mathbf{B}_1 + \cdots + d_m\mathbf{B}_m = \mathbf{B}\mathbf{D}_q \tag{5}$$

where $\mathbf{D}_q$ is an m by 1 column vector of the coordinates of $\mathbf{A}_q$ with respect to the columns of $\mathbf{B}$. Since a current tableau in the Simplex Method came by pivoting from the initial tableau, the coordinates $\mathbf{D}_q$, which are unique, must correspond to the qth column of the current tableau. From (5)

$$\mathbf{D}_q = \mathbf{B}^{-1}\mathbf{A}_q$$

where $\mathbf{D}_q$ is the qth column of the current tableau. This completes the proof of (4) as $\mathbf{D}_q$ is $\overline{\mathbf{A}}_q$.

The elementary matrix $\mathbf{T}$ in (3) may be found in each iteration by the following steps:

ALGORITHM FOR MATRIX T

1. Find the pivot position (p, q) for the next pivot. Any column in the new tableau may be found by (4), so in particular the objective row is determined by multiplying each column of $\mathbf{A}$ by the bottom row of $\overline{\mathbf{B}}^{-1}$ where the current $\overline{\mathbf{B}}^{-1}$ is found from (3). As usual the pivotal column corresponds to the most negative entry in the objective row. The pivotal row is determined by the minimum θ ratio that is nonnegative. Only the pivotal column and the right-hand column are needed to get the θ ratios.

2. Transform the pivotal column, $\overline{\mathbf{A}}_q$, by steps 1 and 3 of the Condensed Tableau Pivoting Algorithm. That is, replace the pivot with its reciprocal

and divide the remaining entries of the column by the negative of the pivot.

3. Replace the pth column of the identity $\mathbf{I}$ with the column vector determined in step 2. This is the matrix $\mathbf{T}$.

The nonunit column vector in $\mathbf{T}$ is called the *eta vector* in the language of modern L.P. codes. As soon as $\mathbf{T}$ is determined, the next B.F.S. may be found by equations (2) and (3)

$$\mathbf{X}_B = \overline{\mathbf{B}}^{-1} \mathbf{C} = \mathbf{T}\,\mathbf{B}^{-1}\,\mathbf{C}.$$

If the B.F.S. at any stage is denoted by $\overline{\mathbf{C}} = \mathbf{B}^{-1}\,\mathbf{C}$, then the new B.F.S. is

$$\mathbf{X}_B = \mathbf{T}\,\overline{\mathbf{C}}.$$

Working from iteration to iteration, the new B.F.S. is matrix $\mathbf{T}$ times the previous solution.

The entire process is illustrated in the following example from section 4.2.

Example 1.

Find $X_1 \geq 0$, $X_2 \geq 0$ such that

$$X_1 + 3X_2 \leq 7$$
$$2X_1 + X_2 \leq 4$$

and $X_1 + 2X_2$ has a maximum value.

The initial extended tableau was given earlier where X_3, X_4 are the slack variables and X_5 is the objective value. The initial B.F.S. is

$$\mathbf{X}_B = \mathbf{B}^{-1}\,\mathbf{C} = \begin{pmatrix} 1 & 0 & 0 \\ 0 & 1 & 0 \\ 0 & 0 & 1 \end{pmatrix} \begin{pmatrix} 7 \\ 4 \\ 0 \end{pmatrix} = \begin{pmatrix} 7 \\ 4 \\ 0 \end{pmatrix}$$

or $X_3 = 7$, $X_4 = 4$, $X_5 = M = 0$. By the Simplex Algorithm the pivot position $(p, q) = (1, 2)$ so X_2 in the second column is the incoming variable and X_3 in the first row goes out of the basis. To find $\mathbf{T}$, according to step 2 of the algorithm, transform $\mathbf{A}_2$ as a pivotal column. Then, following step 3, place the transformed column into the first column of $\mathbf{I}$. The result is

$$\mathbf{T} = \begin{pmatrix} 1/3 & 0 & 0 \\ -1/3 & 1 & 0 \\ 2/3 & 0 & 1 \end{pmatrix}$$

The updated inverse of **B** and the second B.F.S. are

$$\overline{\mathbf{B}}^{-1} = \mathbf{T}\,\mathbf{B}^{-1} = \mathbf{T}$$

$$\mathbf{X}_B = \mathbf{T}\,\mathbf{C} = \begin{pmatrix} 1/3 & 0 & 0 \\ -1/3 & 1 & 0 \\ 2/3 & 0 & 1 \end{pmatrix} \begin{pmatrix} 7 \\ 4 \\ 0 \end{pmatrix} = \begin{pmatrix} 7/3 \\ 5/3 \\ 14/3 \end{pmatrix}$$

or $X_2 = 7/3$, $X_4 = 5/3$, $X_5 = M = 14/3$.

Now repeat the algorithm to find the next update of matrix **T**. The new pivot position is arrived at by following step 1. Take the vector scalar or dot product of the last row of $\mathbf{B}^{-1}$ with each of the first $n-1$ columns of **A** to get the objective coefficients:

$$\overline{a}_{31} = (\,2/3 \quad 0 \quad 1\,) \begin{pmatrix} 1 \\ 2 \\ -1 \end{pmatrix} = -1/3$$

$$\overline{a}_{32} = (\,2/3 \quad 0 \quad 1\,) \begin{pmatrix} 3 \\ 1 \\ -2 \end{pmatrix} = 0$$

$$\overline{a}_{33} = (\,2/3 \quad 0 \quad 1\,) \begin{pmatrix} 1 \\ 0 \\ 0 \end{pmatrix} = 2/3$$

$$\overline{a}_{34} = (\,2/3 \quad 0 \quad 1\,) \begin{pmatrix} 0 \\ 1 \\ 0 \end{pmatrix} = 0$$

The negative in $\overline{a}_{31}$ determines the first column as the pivotal column so X_1 comes into the basis. In order to compute the θ ratios, find the pivotal column in the second tableau by

$$\bar{\mathbf{A}}_1 = \bar{\mathbf{B}}^{-1}\mathbf{A}_1 = \begin{pmatrix} 1/3 & 0 & 0 \\ -1/3 & 1 & 0 \\ 2/3 & 0 & 1 \end{pmatrix} \begin{pmatrix} 1 \\ 2 \\ -1 \end{pmatrix} = \begin{pmatrix} 1/3 \\ 5/3 \\ -1/3 \end{pmatrix}$$

The right-hand column has already been computed in the second B.F.S., so the θ ratios are the quotients

$$\theta_1 = 7/3 \div 1/3 = 7$$

$$\theta_2 = 5/3 \div 5/3 = 1.$$

The last row should be ignored in computing θ ratios since it is the objective row. The minimum ratio is in the second row so that is the pivotal row and X_4 now in the second row goes out of the basis. Thus $(p, q) = (2, 1)$. By algorithm steps 2 and 3 the updated $\mathbf{T}$ is

$$\mathbf{T} = \begin{pmatrix} 1 & -1/5 & 0 \\ 0 & 3/5 & 0 \\ 0 & 1/5 & 1 \end{pmatrix}$$

where the second column of $\mathbf{I}$ has been replaced with a pivot of column $\bar{\mathbf{A}}_1$ just found. The new basis inverse is

$$\bar{\mathbf{B}}^{-1} = \mathbf{T}\,\mathbf{B}^{-1} = \begin{pmatrix} 1 & -1/5 & 0 \\ 0 & 3/5 & 0 \\ 0 & 1/5 & 1 \end{pmatrix} \begin{pmatrix} 1/3 & 0 & 0 \\ -1/3 & 1 & 0 \\ 2/3 & 0 & 1 \end{pmatrix} = \begin{pmatrix} 2/5 & -1/5 & 0 \\ -1/5 & 3/5 & 0 \\ 3/5 & 1/5 & 1 \end{pmatrix}$$

The third B.F.S. follows from

$$\mathbf{X}_B = \bar{\mathbf{B}}^{-1}\mathbf{C} = \begin{pmatrix} 2/5 & -1/5 & 0 \\ -1/5 & 3/5 & 0 \\ 3/5 & 1/5 & 1 \end{pmatrix} \begin{pmatrix} 7 \\ 4 \\ 0 \end{pmatrix} = \begin{pmatrix} 2 \\ 1 \\ 5 \end{pmatrix}$$

or $X_2 = 2, X_1 = 1, X_5 = M = 5$. Note that when X_1 came into the basis, it replaced X_4 that was in the second row. It is even easier to get the new B.F.S. directly from $\mathbf{T}$ by using the previous solution.

$$\mathbf{X}_B = \mathbf{T}\bar{\mathbf{C}} = \begin{pmatrix} 1 & -1/5 & 0 \\ 0 & 3/5 & 0 \\ 0 & 1/5 & 1 \end{pmatrix} \begin{pmatrix} 7/3 \\ 5/3 \\ 14/3 \end{pmatrix} = \begin{pmatrix} 2 \\ 1 \\ 5 \end{pmatrix}$$

The objective row in the third tableau is the set of products:

$$\overline{a}_{31} = (\,^3/_5\ ^1/_5\ 1\,) \begin{pmatrix} 1 \\ 2 \\ -1 \end{pmatrix} = 0$$

$$\overline{a}_{32} = (\,^3/_5\ ^1/_5\ 1\,) \begin{pmatrix} 3 \\ 1 \\ -2 \end{pmatrix} = 0$$

$$\overline{a}_{33} = (.\,^3/_5\ ^1/_5\ 1\,) \begin{pmatrix} 1 \\ 0 \\ 0 \end{pmatrix} = {}^3/_5$$

$$\overline{a}_{34} = (\,^3/_5\ ^1/_5\ 1\,) \begin{pmatrix} 0 \\ 1 \\ 0 \end{pmatrix} = {}^1/_5$$

Since all of the objective coefficients are now positive, the third B.F.S. is optimal. The answer agrees with the answer in Chapter 4 and a close examination of the pivoting in that chapter shows that the computations are equivalent to the matrix algebra.   ●

In addition to the elegance of the matrix notation, the Revised Simplex Method has worthwhile advantages in large problems. Especially in problems with many more columns than rows, a significant savings occurs in the random access storage space needed for a computer to carry the Simplex tableau. At each iteration only two of the tableau columns must be carried in high speed memory along with the objective row and a representation of the **T** matrices. Each **T** matrix is completely determined by the eta vector and a pointer to tell the position of the eta vector in that **T**. The remaining columns are all unit vectors and need not be stored. After any number of iterations the Simplex result may be found by multiplying the string of **T** matrices times any desired column of the original tableau. In the case of a 10 by 1000 tableau, each **T** is only 10 by 10 and can be represented by 11 numbers, the eta vector, and its pointer. On the other hand, carrying the entire tableau would require 10,000 numbers in fast memory.

Another important feature of the Revised Simplex Method is the comparatively high computer speed of these matrix products for sparse matrices. A sparse matrix has a large portion of its entries zero, i.e., the nonzero elements are strikingly few in number. Matrix **T** is generally sparse, and on large applied problems it is common for the **A** matrix to be sparse. The greater the sparsity, the faster these matrix products run.

If **A** is sparse, the columns of **A** may be kept in the fast memory by recording only the nonzero entries with a pointer to indicate their position in the columns.

A final simplification can be made. After many iterations some of the **T** matrices may be redundant, and the newly generated **T**s become less sparse as the eta vectors fill up. By backtracking and constructing only those **T**s necessary to represent a particular known basis, the redundancy may be eliminated. In this way the number of **T** matrices may be limited to the number of columns in the basis.

If a tableau is dense and relatively square, then our Condensed Tableau pivoting is just as rapid and convenient as the Revised Simplex Method.

PROBLEMS

7-1. For the following matrix game

$$\begin{bmatrix} 1 & 4 & 5 \\ 7 & 2 & 6 \\ 8 & 9 & 3 \end{bmatrix}$$

(a) Set up the dual linear programs for A's problem and B's problem.
(b) Pivot the condensed tableau for B and pivot by the Dual Simplex Method the condensed tableau for A, verifying at each pivot the Fundamental Duality Theorem.
(c) Verify von Neumann's Minimax Theorem.
(d) State the optimal strategies and the value of the game.

7-2. Follow the same procedure in problem 7-1 for the following matrix game.

$$\begin{bmatrix} 1 & -2 & 3 \\ -4 & 5 & 6 \end{bmatrix}$$

7-3. Solve the following linear program by first dualizing the problem and then pivoting. Carefully read the final tableau to obtain the complete solution vector. That is, state the values of the four slack variables as well as the three main variables of the original problem. As a check on your result convert the type II inequalities into type I, and then pivot by the Dual Simplex Algorithm.

$$x_i \geq 0, \ i = 1, 2, 3$$
$$x_1 + 2x_2 \qquad \geq 2$$
$$3x_1 + \ x_2 + \ x_3 \geq 4$$
$$4x_3 \geq 1$$
$$x_1 \qquad + 3x_3 \geq 1$$

where $4x_1 + 3x_2 + 3x_3 = m$ is a minimum.

7-4. For the following linear program carry out the same directions given in problem 7-3.

$$x_i \geq 0, \, i = 1, 2, 3$$
$$x_1 + 2x_2 \qquad \geq 2$$
$$3x_1 + \; x_2 + \; x_3 \geq 4$$
$$4x_3 \geq 1$$
$$2x_1 \qquad + \; x_3 \geq 1$$

where $4x_1 + 3x_2 + 3x_3 = m$ is a minimum.

7–5. Solve the following linear program and state the complete solution vector including the five slack variables.

$$x_1 \geq 0, x_2 \geq 0$$
$$3x_1 + 2x_2 \geq 4$$
$$2x_1 + \; x_2 \geq 3$$
$$-x_1 + 3x_2 \geq 5$$
$$-2x_1 + \; x_2 \geq 2$$
$$4x_1 + 2x_2 \geq 5$$

where $x_1 + x_2 = m$, a minimum.

7–6. Solve the following linear program and state the complete solution vector including all slack variables.

$$x_i \geq 0, \, i = 1, 2, 3$$
$$x_1 + 2x_2 \qquad \geq 4$$
$$3x_1 + \; x_2 + \; x_3 \geq 5$$
$$x_2 + 4x_3 \geq 2$$
$$x_1 \qquad + \; x_3 \geq 3$$

where $4x_1 + 3x_2 + 3x_3 = m$, a minimum.

7–7. Solve the following L.P. for the unknowns and find the minimum.

$$x_i \geq 0, \, i = 1, 2, 3$$
$$3x_1 + \; 2x_2 + \; 4x_3 \geq 160$$
$$4x_1 + \; \; x_2 + \; 5x_3 \geq 210$$
$$2x_1 + \; 2x_2 + \; 4x_3 \geq 100$$
$$57x_1 + 27x_2 + 73x_3 = m, \text{ a minimum.}$$

7–8. Dualize the problem in 7–7 and solve for the dual variables, that is, find the optimal values of the new variables in the dual constraints along with the objective maximum. Does the maximum equal the minimum of problem 7–7?

7–9. Solve the following relaxation of a knapsack problem. Knapsack problems will be taken up in Chapter 11 where the answers are forced to be integer.

$$x_i \geq 0, \, i = 1, 2, 3$$
$$4x_1 + \; 3x_2 + \; 6x_3 \leq 6$$
$$x_1 \qquad\qquad \leq 1$$
$$x_2 \qquad \leq 1$$
$$x_3 \leq 1$$
$$30x_1 + 18x_2 + 40x_3 = M, \text{ a maximum.}$$

7-10. Dualize the problem in 7–9 and find the optimal values of the dual variables along with the objective minimum. Can the solution to both problems be read off of the optimal tableau of either one?

7-11. In a card matching game, player A has three cards, a red 2, a black 2, and a red 3. Player B also has three cards, a red 2, a black 2, and a black 3. A and B match their cards with the following rules: If both cards are red or both cards black, then A wins the face value of his card. If both cards are 3's, then the game is a tie. Otherwise, for any combination of red and black, B wins the face value of his card. What is the game value, and what are the optimal strategies of players A and B?

7-12. In problem 7–11, change the deuces to tens. In other words, replace the red 2's with red 10's and the black 2's with black 10's. Now what is the game value and the optimal strategies? Is the game fair?

Solve the following problems by the Revised Simplex Method.

7-13. Maximize M such that:
$$x_i \geq 0, \quad i = 1, 2$$
$$x_1 + 3x_2 \leq 10$$
$$5x_1 + 4x_2 \leq 28$$
$$4x_1 + 5x_2 = M.$$

7-14. Maximize M such that:
$$x_i \geq 0, \quad i = 1, 2, 3$$
$$4x_1 + 3x_2 + x_3 \leq 46$$
$$3x_1 + 6x_2 + 2x_3 \leq 47$$
$$12x_1 + 5x_2 + 6x_3 = M.$$

7-15. Find the solution vector and the maximum value of M such that:
$$x_i \geq 0, \quad i = 1, 2$$
$$5x_1 + 2x_2 \leq 1000$$
$$5x_1 + 4x_2 \leq 1300$$
$$x_1 + 2x_2 \leq 500$$
$$2x_1 + x_2 = M.$$
Check your answers graphically.

7-16. Repeat problem 7–15 with the objective function $M = 3x_1 + 4x_2$.

7-17. From Chapter 2 solve problem 2–14 for the Seeall Company.

7-18. Solve problem 2–17 for the Neely Nut Company.

7-19. Solve problem 4–8 for Caroline's Quality Candy.

7-20. Classic Container has an order to fill on two sizes of cardboard boxes. The smaller box uses 1½ square feet of cardboard and sells for a profit of 28 cents per box. The larger box uses 3 square feet of cardboard and sells for a profit of 57 cents per box. The customer will take no more than 2500 of the smaller boxes on this order. Classic's machinery must make at least twice as many small boxes as large boxes in the allotted time. The inventory shows that 6000 square feet of the appropriate cardboard is on hand for this project. If all of the boxes are sold, how many of each size should Classic Container produce, and what is their profit from this sale?

7-21. Suppose extra handling on the larger box of problem 7-20 reduces its profit by 8 cents per box. Then what should the optimum solution and profit be for Classic Container?

CHAPTER 8

Primal-Dual Methods

8.1 METHODS FOR MIXED SYSTEMS

The concept of duality not only has great theoretical significance but it also has considerable value in solving problems. The dual of a particular program may be much easier to solve than the original program. By a combination of the Simplex Algorithm and the Dual Simplex Algorithm, artificial variables may be avoided completely. Avoiding artificial variables reduces the size of the tableau and usually makes a marked reduction in the number of iterations necessary to optimize the problem.

Let us consider the following example in two variables which may be graphed, and then the various solutions may be checked against the graph.

Example 1. Find $x_1 \geq 0$, $x_2 \geq 0$, satisfying constraints

$$x_1 + 7x_2 \leq 63 \tag{1}$$

$$3x_1 + x_2 \geq 9 \tag{2}$$

$$3x_1 + 2x_2 \geq 15 \tag{3}$$

$$5x_1 + 6x_2 \geq 30 \qquad\qquad\qquad (4)$$

$$x_1 + 4x_2 \geq 8 \qquad\qquad\qquad (5)$$

$$8x_1 + 3x_2 \leq 80, \qquad\qquad\qquad (6)$$

such that $2x_1 + x_2$ is a minimum.

The graph in Figure 8–1 shows the minimum at vertex (1, 6). The boundaries are labeled with the numbers of their corresponding constraints.

The given system of constraints contains four type II inequalities. Under the Simplex Method these will require four artificial variables. As an exercise, solve the problem with artificial variables. Meanwhile let us label the six slack variables $x_3, \ldots, x_8$, multiply constraints (1) and (6) by -1, and then dualize. The symmetric system is dualized by definition 10 in Chapter 6 to give the following primal program.

Find $x_i \geq 0$, $i = 3, \ldots, 8$ satisfying

$$-1x_3 + 3x_4 + 3x_5 + 5x_6 + 1x_7 - 8x_8 \leq 2 \qquad\qquad (7)$$

$$-7x_3 + 1x_4 + 2x_5 + 6x_6 + 4x_7 - 3x_8 \leq 1 \qquad\qquad (8)$$

$$-63x_3 + 9x_4 + 15x_5 + 30x_6 + 8x_7 - 80x_8 = \overline{M},$$

where $\overline{M}$ is a maximum.

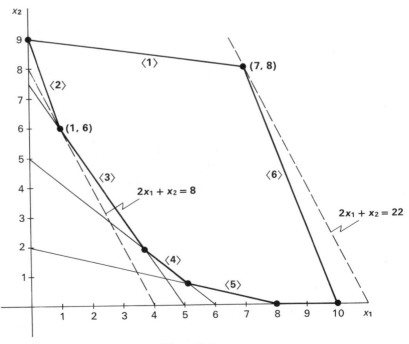

Figure 8–1

The von Neumann Theorem tells us that $\overline{M} = m$, the minimum of the dual program. Note that we are considering the problem as originally stated to be the dual. We therefore solve the primal program for its maximum by using the initial tableau of constraints (7) and (8) where x_1 and x_2 are the slack variables.

	3	4	5	6	7	8	
1	−1	3	3	5	1	−8	2
2	−7	1	2	⑥	4	−3	1
	63	−9	−15	−30	−8	80	0

The tableau may be run on our automatic routine from Chapter 5, which produces the final tableau listed next after three pivots.

	3	1	6	2	7	8	
4	6.33	.67	−2.67	−1	−3.33	−2.33	.33
5	−6.67	−.33	4.33	1	3.67	−.33	.33
	20	1	11	6	17	54	8

It is very important to read the final tableau correctly. Because we want the solution to the dual program, the basic variables are across the top of the final tableau with their corresponding values in the bottom row. The nonbasic variables are at the left. Thus the complete solution vector, including the slack variables, is

$x_1 = 1$

$x_2 = 6$

$x_3 = 20$

$x_4 = 0$

$x_5 = 0$

$x_6 = 11$

$x_7 = 17$

$x_8 = 54.$

The minimum value 8, in the lower right-hand corner, checks the value found from the objective function,

$$m = 2x_1 + x_2 = 2 + 6 = 8.$$

Note that x_4 and x_5 are zero since they are nonbasic. Only three pivots were required compared to six pivots to optimize the tableau with artificial variables. Also the tableau size has been reduced from 8 × 7 to 3 × 7.

An alternative to dualizing is to use the Dual Simplex Algorithm. Going back to the original constraints of example 1, we multiply the four type II constraints by minus 1. Then setting $m = -M$ will put the problem into the following form.

Maximize M subject to

$$x_1 \geq 0, \ x_2 \geq 0$$
$$x_1 + \ 7x_2 \leq 63$$
$$-3x_1 - \quad x_2 \leq -9$$
$$-3x_1 - \ 2x_2 \leq -15$$
$$-5x_1 - \ 6x_2 \leq -30$$
$$-x_1 - \ 4x_2 \leq -8$$
$$8x_1 + \ 3x_2 \leq 80$$
$$2x_1 + \quad x_2 + M = 0.$$

The original example has now been changed to a program with type I constraints for which the initial tableau is

	1	2	
3	1	7	63
4	−3	−1	−9
5	−3	−2	−15
6	−5	(−6)	−30
7	−1	−4	−8
8	8	3	80
	2	1	0

This tableau is precisely the dual of the previous initial tableau. It is optimal but not feasible. If we pivot by the Dual Simplex Algorithm, the three pivots will be dual to the same three pivots of the dual tableau. For example, the θ ratios of the fourth row are $\theta_1 = -2/5$ and $\theta_2 = -1/6$. Since the numerically smallest ratio is θ_2, the first pivot will occur at (4, 2), fourth row, second column. The initial pivot in our previous solution was at (2, 4). The result after three pivots by the Dual Simplex Algorithm is the final tableau given below.

	4	5	
3	−6.33	6.67	20
1	− .67	.33	1
6	2.67	−4.33	11
2	1	−1	6
7	3.33	−3.67	17
8	2.33	.33	54
	.33	.33	−8

The two final tableaux are of course dual to one another. In this case $M = -8$, so the minimum $m = -M = 8$. ●

The two methods are equivalent, but the first had the advantage of running on our automatic program of Chapter 5. The second must either be programmed by the Dual Simplex Algorithm or have the pivot position told to the machine after each tableau.

Let us now find the maximum of the same objective function over the given feasible region of example 1.

Example 2.   From Figure 8–1, the maximum is seen to occur at vertex (7, 8) and thus $M = 2x_1 + x_2 = 22$. To find it by pivoting, we set up the following initial tableau which differs from the last initial tableau only in the bottom row.

	1	2	
3	1	7	63
4	−3	−1	−9
5	−3	−2	−15
6	−5	−6	−30
7	−1	−4	−8
8	⑧	3	80
	−2	−1	0

The present tableau is neither optimal nor feasible. The difficulty can be handled by a combination of our two methods. All possible pivots should be considered by both the Simplex and Dual Simplex Algorithms. In this case the Dual Simplex Algorithm produces no pivot since the last row is all negative. There are no possible negative θ ratios for determining a pivotal column. For the first column the Simplex Algorithm gives us the row ratios $\theta_1 = 63/1$ and $\theta_6 =$

80/8. The smaller ratio, θ_6, determines the pivot position to be (6, 1). After pivoting we have the following tableau.

	8	2	
3	−.125	(6.625)	53
4	.375	.125	21
5	.375	− .875	15
6	.625	−4.125	20
7	.125	−3.625	2
1	.125	.375	10
	.25	−.25	20

The second tableau is already feasible but not optimal. The next pivot determined by row ratios using the second column is at (1, 2). This leads to the final tableau below.

	8	3	
2	−.019	.151	8
4	.377	−.019	20
5	.358	.132	22
6	.547	.623	53
7	.057	.547	31
1	.132	−.057	7
	.245	.038	22

The maximum is 22 and the complete solution vector is:

$x_1 = 7$

$x_2 = 8$

$x_3 = 0$

$x_4 = 20$

$x_5 = 22$

$x_6 = 53$

$x_7 = 31$

$x_8 = 0.$ ●

A program using artificial variables will find the same result in seven pivots compared to the two pivots used here. Also the tableau size is cut down from 8×7 to 7×3. So far we have had no conflicts in choosing a pivot. The possibility remains that a tableau is neither feasible nor optimal and the two methods each produce different pivots. A choice may be made by answering the question, "Which pivot makes the greater change in the objective value?" A pivot chosen by the Simplex Algorithm increases the objective value, advancing toward optimality. A pivot chosen by the Dual Simplex Algorithm decreases the objective value, which both improves the objective of the dual and advances the primal toward feasibility.

To distinguish between the two types of θ ratios, let us use θ_i, $i = 1$ to $m - 1$, for the row ratios and $\bar{\theta}_j$, $j = 1$ to $n - 1$, for the column ratios. From Chapter 4 the formula for the objective value after a pivot at Y_{pq} is

$$\bar{Y}_{mn} = Y_{mn} - Y_{mq}(Y_{pn}/Y_{pq})$$
$$= Y_{mn} - Y_{mq}\theta_p.$$

The increase in the objective function is the absolute value of $Y_{mq}\theta_p$. Let

$$I_p = | Y_{mq}\theta_p| \qquad (9)$$

be the increase in the objective function due to a pivot chosen by the use of row ratios. The corresponding formula for the decrease due to a pivot chosen by column ratios is

$$Y_{mq} Y_{pn}/Y_{pq} = \bar{\theta}_q Y_{pn}.$$

Let

$$D_q = |\bar{\theta}_q Y_{pn}| \qquad (10)$$

be the decrease in the objective function due to a pivot chosen by the Dual Simplex Algorithm. When two pivots are possible the choice may be made by picking the pivot corresponding to the larger of I_p and D_q. Thus the next tableau will make the largest possible change in the program either toward optimality or toward feasibility. In most cases this choice should result in the fewest number of pivots necessary to reach a final tableau.

Let us solve the following example.

Example 3. Find $x_i \geq 0$, $i = 1, 2, 3, 4$, satisfying the constraints

$$x_1 - 2x_2 + 4x_3 - 3x_4 = 10$$
$$2x_1 + 3x_2 - 1x_3 + 5x_4 \leq 15$$
$$3x_1 - 1x_2 + 2x_3 + 3x_4 \geq 12,$$

where $x_1 + x_2 - 2x_3 - 4x_4 = M$ is a maximum.

The equality constraint is replaced with two inequalities, one of type I and one of type II. The type II inequalities are then multiplied by -1 to form a symmetrical maximizing problem. The equivalent system is

$$x_1 - 2x_2 + 4x_3 - 3x_4 \le 10$$
$$-x_1 + 2x_2 - 4x_3 + 3x_4 \le -10$$
$$2x_1 + 3x_2 - x_3 + 5x_4 \le 15$$
$$-3x_1 + x_2 - 2x_3 - 3x_4 \le -12$$
$$M - x_1 - x_2 + 2x_3 + 4x_4 = 0.$$

The slack variables are taken to be $x_5, x_6, x_7,$ and x_8, respectively, in the following initial tableau.

	1	2	3	4	
5	1	−2	4	−3	10
6	−1	2	−4	3	−10
7	2	3	−1	5	15
8	−3	1	⊝2	−3	−12
9	−1	−1	2	4	0

The initial tableau is neither feasible nor optimal. The possible row ratios for the first column are $\theta_1 = 10/1$ and $\theta_3 = 15/2$. Using the smaller ratio we compute

$$I_p = I_3 = |\ Y_{51}\theta_3\ | = 7.5.$$

The possible column ratios for the fourth row are $\bar{\theta}_3 = 2/{-2}$ and $\theta_4 = 4/{-3}$. Choosing the smaller ratio in absolute value, we find

$$D_q = D_3 = |\ \bar{\theta}_3\ Y_{45}\ | = 12.$$

Since D_3 is larger than I_3 we pick the pivot according to the Dual Simplex Algorithm to be at (4, 3). The result of this pivot is to decrease M as shown in the next tableau.

	1	2	8	4	
5	−5	0	2	−9	−14
6	⑤	0	−2	9	14
7	3.5	2.5	−.5	6.5	21
3	1.5	−.5	−.5	1.5	6
9	−4	0	1	1	−12

The second tableau is still both infeasible and nonoptimal. We have reduced the infeasible variables to one even though $M = -12$ is further from a maximum. Once again, computing the row ratios for the first column and the column ratios for the first row, we find

$$I_2 = \mid Y_{51}\theta_2 \mid = 4(\; ^{14}\!/_5\;) = 11.2$$

$$D_4 = \mid \bar{\theta}_4 Y_{15} \mid = (\; ^1\!/_9\;)14 = 1.56.$$

Since I_2 is the larger we pivot at (2, 1) to obtain the next tableau.

	6	2	8	4	
5	1	0	0	0	0
1	.2	0	−.4	1.8	2.8
7	−.7	2.5	ⓐ.9	.2	11.2
3	−.3	−.5	.1	−1.2	1.8
9	.8	0	−.6	8.2	−.8

The third tableau is feasible but not yet optimal. The objective value has been increased to $M = -.8$. A final pivot is determined by the row ratios of the third column to be at (3, 3). Such a pivot must increase the objective value as shown in the final tableau.

	6	2	7	4	
5	1	0	0	0	0
1	−.111	1.111	.444	1.889	7.778
8	−.778	2.778	1.111	.222	12.444
3	−.222	−.778	−.111	−1.222	.556
9	.333	1.677	.667	8.333	6.667

The solution vector for the four main variables is:

$x_1 = 7.778$

$x_2 = 0$

$x_3 = .556$

$x_4 = 0.$

For this solution the maximum of the objective function is $M = 6.667$ as may be checked. Slack variables x_5 and x_6 are necessarily both zero since they arose from an equality constraint. The final tableau appears to be degenerate with a basic variable equal to zero. However, no further pivoting is possible and so the solution is unique. Slack variables associated with an equality constraint must be zero in the final tableau and do not represent a true degeneracy in the optimal basic solution.   •

The solution to this example may be achieved in four pivots by use of artificial variables.

8.2   THE PRIMAL-DUAL ALGORITHM

In each previous example a worthwhile savings was realized by the combined methods both in the number of iterations necessary and in tableau size. Let us summarize our technique in algorithm form.

THE PRIMAL-DUAL ALGORITHM FOR MIXED SYSTEMS

1. Replace equality constraints with two inequalities, one of type I and the other of type II.

2. Convert all type II inequalities to type I by multiplying by -1.

3. If necessary, convert the objective function to a maximization objective.

4. Set up the initial tableau with a slack variable in the basis for each constraint. Eventually all variables are to satisfy the nonnegativity requirement.

5. For the most negative number in the last column, determine a possible pivot in that row by column ratios according to the Dual Simplex Algorithm. The ratio must be negative. If no pivot is available, move to the row of the next most negative number in the last column and try again. Continue through the negatives of the last column until either a possible pivot is found or none is available.

6. For the most negative number in the bottom row, determine a possible pivot in that column by row ratios as in the Simplex Algorithm. The ratio must be positive. If no pivot is available, continue to try the remaining columns with a negative in the bottom row until either a pivot is found or none is available.

7. If steps 5 and 6 each produce a pivot, compute I_p and D_q by formulas (9) and (10) respectively in section 8.1. Choose the pivot corresponding to the larger, I_p or D_q.

8. After pivoting repeat steps 5 through 7 until the tableau is optimal and feasible or until no new pivot can be found.

A correct interpretation of the final tableau is of prime importance. The final tableau may be optimal and feasible. In this case, a solution has been found to both the primal and dual programs. The solution may be degenerate, but true degeneracy should be distinguished from that arising when equality constraints are present. When an equality constraint is satisfied, the two asso-ciated slack variables are reduced to zero. One of these slacks is driven out of the basis, but the other remains in the solution at value zero. This is not a true degeneracy as was noted in our last example. If the final tableau is optimal but contains a negative in the last column due to the fact that no pivot can be found in that row, then the primal program has no feasible solutions and its dual is unbounded. If the final tableau is feasible but contains a negative in the objective row for which all other entries in that column are nonpositive, then the primal objective is unbounded and the dual program has no feasible solutions. The final tableau may also be both infeasible and nonoptimal. For an example see problem 22. One of these situations will necessarily occur so that the original problem has an optimal solution, or it is unbounded, or it is inconsistent.

The possible outcomes for a primal problem and its dual are summar-ized in the following table.

PRIMAL (MAXIMUM)	DUAL (MINIMUM)
Optimum solution with objective $\overline{M}$	Optimum solution with objective $m = \overline{M}$
Unbounded solution	No solution (inconsistent)
No solution (inconsistent)	Unbounded solution
No solution (inconsistent)	No solution (inconsistent)

THEOREM 8-1. *If both the primal and dual problems have feasible solutions, then an optimal feasible solution exists for both problems and the two objective values are equal.*

In reading the answers from the final tableau, care must be taken to distinguish between the basic and nonbasic variables. For the primal program the basic variables are denoted by the subscripts $L(I)$ to the left of our tableau, and their values are found in the right-hand column. For the dual program the basic variables are denoted by the subscripts $K(J)$ across the top of our tableau, and their corresponding values are found in the bottom row. In both cases the nonbasic variables are always zero.

A set of subroutines in BASIC and FORTRAN languages for carrying out the Primal-Dual Algorithm is found in Appendix A.

8.3 COMPLEMENTARY SLACKNESS

Another important observation, which is readily obtained from our primal-dual tableau, is called *complementary slackness* in dual linear programs. Sup-pose a pair of dual programs have optimal feasible solutions. If there is positive slack in a constraint of one program, then the corresponding dual variable is zero. Also if one of the main variables in one program is basic in the solution, then the slack in the corresponding dual constraint is zero. Both these facts result from our definition of duality that interchanges the slack variables of one program with the main variables of its dual. When the final tableau is reached with its optimal feasible solutions, a basic variable for one problem is nonbasic for its dual. So whenever one is positive, its dual correspondent is zero. One way of stating the conclusion is

Theorem 8-2. *(Theorem of Complementary Slackness) Any pair of dual optimal feasible solutions satisfy the following:*

1. If the ith slack variable of the primal is basic, then the corresponding i^{th} dual variable is zero.
2. If the jth main variable of the primal is basic, then the corresponding j^{th} slack variable of the dual is zero.

In a degenerate case both a basic variable and its dual correspondent can be zero. For a pair of nondegenerate programs where all basic variables are positive, the following chart pairs up the possibilities.

PRIMAL	DUAL
(1) $a_1x_1 + a_2x_2 + a_3x_3 \leq a_4$	(1) $a_1x_4 + b_1x_5 \geq c_1$
(2) $b_1x_1 + b_2x_2 + b_3x_3 \leq b_4$	(2) $a_2x_4 + b_2x_5 \geq c_2$
$c_1x_1 + c_2x_2 + c_3x_3 = M$	(3) $a_3x_4 + b_3x_5 \geq c_3$
	$a_4x_4 + b_4x_5 = m$
slack variables x_4, x_5	slack variables x_1, x_2, x_3

$x_4 > 0$, thus constraint (1) is $<$	$x_4 = 0$, out of basis
$x_5 > 0$, thus constraint (2) is $<$	$x_5 = 0$, out of basis
$x_1 > 0$	$x_1 = 0$, thus constraint (1) is $=$
$x_2 > 0$	$x_2 = 0$, thus constraint (2) is $=$
$x_3 > 0$	$x_3 = 0$, thus constraint (3) is $=$
$x_4 = 0$, thus constraint (1) is $=$	$x_4 > 0$ in basis
$x_5 = 0$, thus constraint (2) is $=$	$x_5 > 0$ in basis
$x_1 = 0$, out of basis	$x_1 > 0$, thus constraint (1) is $>$
$x_2 = 0$, out of basis	$x_2 > 0$, thus constraint (2) is $>$
$x_3 = 0$, out of basis	$x_3 > 0$, thus constraint (3) is $>$

Each line of the table may be read with an implication going either way for the nondegenerate programs. The concept of complementary slackness has economic meaning as seen in the next section on shadow prices.

8.4 SHADOW PRICES

One of the important interpretations of duality is discovered in the economic meaning of the dual solution. For example, a manufacturer is interested in knowing the highest price he can afford to pay for additional material to increase his output. The question of whether or not it is profitable to buy raw material at a certain cost may be answered by what is called the *shadow price* of that commodity. The idea will be illustrated in the following example.

Example 4. A women's apparel manufacturer has a production line making two styles of girls' skirts. Style one is a skirt which requires 2 ounces of cotton thread, 3 ounces of dacron thread, and 3 ounces of linen thread. Style two is a miniskirt which requires 2 ounces of cotton thread, 2 ounces of dacron thread, and 1 ounce of linen thread. The manufacturer realizes a net profit of $1.95 on style one and a net profit of $1.59 on style two. He has on hand an inventory of 15 pounds of cotton thread, 16 1/4 pounds of dacron thread, and 13 3/4 pounds of linen thread. His immediate problem is to determine a production schedule, given the current inventory, to make a maximum profit. Then he would like to know at what prices per oz. it would be profitable to buy more thread, assuming a small increase in production.

SOLUTION. We will first solve the immediate problem. Let x_1 be the number of skirts produced and x_2 the number of miniskirts produced with the current inventory. Expressing the quantities of thread in ounces, we arrive at the following constraints:

$$2x_1 + 2x_2 \leq 240$$
$$3x_1 + 2x_2 \leq 260$$
$$3x_1 + 1x_2 \leq 220.$$

The objective is to maximize the profit.

$$M = 1.95 \, x_1 + 1.59 \, x_2.$$

Let x_3 be the slack ounces of cotton, x_4 be the slack ounces of dacron, and x_5 be the slack ounces of linen. The initial tableau is

	1	2	
3	2	2	240
4	3	2	260
5	3	1	220
6	−1.95	−1.59	0

After three pivots, we reach the final tableau of the primal program.

		3	4	
5		1.5	−2	60
2		1.5	−1	100
1		−1	1	20
6		.435	.36	198

With the thread on hand, the manufacturer should produce $x_1 = 20$ skirts and $x_2 = 100$ miniskirts at a total profit of \$198. By the Duality Theorem, this optimum value may be computed either from the objective function of the primal program or the objective function of the dual program. These two objective functions are given next in equations (1) and (2) where the x_i's are read off the final tableau.

$$M = 1.95\,x_1 + 1.59\,x_2 \qquad\qquad (1)$$
$$= 1.95\,(\,20\,) + 1.59\,(\,100\,) = \$198.$$

$$m = 240\,x_3 + 260\,x_4 + 220\,x_5 \qquad\qquad (2)$$
$$= 240\,(\,.435\,) + 260\,(\,.36\,) + 220\,(\,0\,) = \$198.$$

In the dual problem, each of the variables must be given an appropriate economic meaning that is quite different from their definition in the primal problem. Variables x_3, x_4, and x_5 may be thought of as the cost per ounce of cotton, dacron, and linen respectively. The dual objective is then to find x_3, x_4, and x_5, so that these unit costs will minimize equation (2). Of course, the production schedule that maximizes profit on skirts is the same as the production schedule that minimizes the total cost of raw material. These unit costs, $x_3 = .435$, $x_4 = .36$, and $x_5 = 0$, also represent an upper bound on the amount the manufacturer should pay for additional raw material. For example, if he buys an additional ounce of cotton, equation (2) shows that his profit is increased by 43 ½ cents. Likewise, an additional ounce of dacron would increase profit by 36 cents. ●

DEFINITION.   The *shadow price* of one unit of a commodity corresponding to a nonbasic variable is a value that is equal to the increase in profit to be realized by adding a unit of that raw material into the problem basis. Another popular term for shadow price is *marginal price*.

Naturally the shadow price is the highest price that the manufacturer would be willing to pay for more raw material. He wishes to buy raw material at less than its shadow price in order that the corresponding increase in output will also increase net profits.

The shadow prices of cotton and dacron from example 4 are 43 ½ cents per ounce and 36 cents per ounce respectively. The shadow price of linen is zero since it does not increase profit to buy more linen until the current supply is exhausted. In the basic solution to the primal program, x_5 shows us that there are 60 ounces of slack or unused linen. In general, if a slack variable is nonzero in the solution, then its shadow price will be zero by the theorem of complementary slackness.

Equation (2) may be used to find the optimum of a new problem without resolving provided the basic variables don't leave the basis. We have already noted in example 4 that if an additional ounce of dacron is available, the new profit would be $198.36. As a check, let us solve example 4 with 261 ounces of dacron. The final tableau, reached in three pivots, is

	3	4	
5	1.5	−2	58
2	1.5	−1	99
1	−1	1	21
6	.435	.36	198.36

The only change is in the last column where the solution requires the production of one more skirt and one less miniskirt for the new profit of $198.36.

The question of how much a resource can be changed while the shadow prices remain the same is a problem in Sensitivity Analysis. Geometrically, this problem may be viewed as how much can the coefficients of the dual objective function be changed without slipping off the optimal vertex in the dual solution? If the dual program can be graphed, the answer is seen in considering the slopes involved. In example 4 the dual program may be stated as follows. Find x_3, the cost per ounce of cotton thread, x_4, the cost per ounce of dacron thread, and x_5, the cost per ounce of linen thread, satisfying

$$2x_3 + 3x_4 + 3x_5 \geq 1.95 \qquad\qquad\qquad (3)$$
$$2x_3 + 2x_4 + x_5 \geq 1.59,$$

such that $240\,x_3 + 260\,x_4 + 220\,x_5 = m$ is the minimum total cost of raw material.

In the dual objective function the ratio of cotton to dacron is $12/13$. The corresponding ratios in the two constraints (3) are $2/3$ and $2/2 = 1$. This suggests that in the objective function, cotton could be increased by 20 ounces to 260 ounces so that its ratio to dacron is one. Up to an increase of 20 ounces of cotton, the objective plane will intersect the same optimal vertex. At this point the solution becomes degenerate, but the optimum is still given by

$$m = 260\,(\,.435\,) + 260\,(\,.36\,) + 0$$
$$= 113.10 + 93.60 = \$206.70.$$

We can verify this minimum by solving example 4 with the resource of cotton at 260 ounces. The result is given in the next tableau reached after three pivots.

	3	4	
5	1.5	−2	90
2	1.5	−1	130
1	−1	1	0
6	.435	.36	206.70

The optimum now requires the production of 130 miniskirts and no skirts.

Suppose only the dacron is increased. The increase is limited to 30 ounces because at that point the slack linen is used up and the solution again becomes degenerate. Let us check the answers for an increase of 29 to 289 ounces of dacron. The solution using our shadow prices is immediately

$$m = 240(\ .435\ ) + 289(\ .36\ ) + 0$$

$$= 104.40 + 104.04 = \$208.44.$$

The verification of this minimum by solving example 4 with the resource of dacron at 289 ounces is seen in the following final tableau.

	3	4	
5	1.5	−2	2
2	1.5	−1	71
1	−1	1	49
6	.435	.36	208.44

From this final tableau the optimum production schedule is 49 skirts and 71 miniskirts. In each of the variations of example 4, the maximum profit was computed from the shadow prices without resorting to the solution of the new program. A variety of initial conditions was possible under the same shadow prices. A further look at Sensitivity Analysis will be made in Chapter 9 with the use of a parameter in the resource column.

PROBLEMS

8–1. Solve the following linear program in several ways considering both primal and dual methods. Check your result graphically. State the optimum value and the entire solution vector including slack variables.

$$x_1 \geq 0, x_2 \geq 0$$
$$2x_1 + x_2 \geq 6$$
$$2x_1 + 3x_2 \geq 10$$
$$-3x_1 + 5x_2 \geq -24$$
$$5x_1 + 2x_2 \leq 71$$
$$2x_1 + 7x_2 \leq 78$$
$$3x_1 - 4x_2 \geq -28,$$

where $5x_1 + 3x_2 = M$ is a maximum.

8-2. Carry out problem 8-1 where the objective is to minimize $5x_1 + 3x_2$.

8-3. Solve the following problem without using artificial variables and note the values of all slack variables.

$$x_1 \geq 0, x_2 \geq 0$$
$$2x_1 + x_2 \leq 35$$
$$x_2 \leq 13$$
$$4x_1 + x_2 \geq 12$$
$$2x_1 + x_2 \geq 10$$
$$x_1 + x_2 \geq 7$$
$$x_1 + 4x_2 \geq 10$$
$$x_1 - x_2 \leq 10,$$

where $3x_1 + x_2 = M$ is a maximum.

8-4. Find the minimum value of the objective function in problem 8-3 along with its solution vector.

8-5. Find both the maximum and the minimum value of $2x_1 - 3x_2$ under the constraints given in problem 8-3 without using artificial variables.

8-6. Solve problem 8-5 with artificial variables and compare the number of pivots needed with the number used in problem 8-5.

8-7. Find the solution vector and the minimum value of m for

$$x_i \geq 0, i = 1, 2, 3$$
$$x_1 + 2x_2 - x_3 \leq 9$$
$$2x_1 - x_2 + 2x_3 = 4$$
$$-x_1 + 2x_2 + 2x_3 \geq 5$$
$$2x_1 + 4x_2 + x_3 = m.$$

8-8. Find nonnegative numbers x_i, $i = 1, \ldots, 5$, and the maximum, M, such that

$$2x_1 + 3x_2 - 4x_3 + 3x_4 - 2x_5 = 2$$
$$3x_1 - 2x_2 + 5x_3 - 4x_4 + 3x_5 = 7$$
$$x_1 + 2x_2 - 3x_3 - 3x_4 + 2x_5 = M.$$

8-9. Does the objective function in problem 8–8 possess a minimum under the same constraints?

8-10. Find nonnegative numbers x_i, $i = 1, 2, 3$, and the minimum m, such that

$$3x_1 + 2x_2 + 4x_3 \le 26$$
$$x_1 + 2x_2 + x_3 \ge 13$$
$$3x_1 + x_2 + 2x_3 = 16$$
$$x_1 + 2x_2 + x_3 = m.$$

8-11. Find the nonnegative numbers x_i, $i = 1, 2, 3$, and the minimum, m, such that

$$3x_1 + 2x_2 + 4x_3 \le 26$$
$$2x_1 + x_2 + 2x_3 \ge 13$$
$$2x_1 + 5x_2 + 3x_3 \ge 17$$
$$3x_1 + x_2 + 2x_3 = 16$$
$$x_1 + 2x_2 + x_3 = m.$$

What are the values of the three slack variables?

8-12. Solve the following problem.

$$x_i \ge 0, \quad i = 1, 2, 3, 4$$
$$5x_1 - 3x_2 + 2x_3 + 4x_4 \le 21$$
$$-6x_1 + 2x_2 - 3x_3 + 3x_4 = 1$$
$$4x_1 + 5x_2 + 6x_3 - 5x_4 \ge 12$$
$$x_1 + x_2 + x_3 + x_4 \le 10,$$

where $x_1 + 2x_2 - 3x_3 + x_4 = M$, a maximum.

8-13. Solve problem 8–12 where the objective function is to be minimized.

8-14. Solve the following problem where x_3 is unrestricted in sign, that is, x_3 may be positive, negative, or zero.

$$x_1 \ge 0, \qquad x_2 \ge 0$$
$$3x_1 + 5x_2 + 4x_3 \le 41$$
$$2x_1 - 6x_2 - 9x_3 \le 11$$
$$-x_1 + 3x_2 + 2x_3 \le 7,$$

where $-9x_1 + 30x_2 - 40x_3 = M$, a maximum.

Hint: Replace x_3 with two variables, say y_3 and y_4, such that $x_3 = y_3 - y_4$ where both y_3 and y_4 are nonnegative. Then x_3 may be negative if $y_3 < y_4$, positive if $y_3 > y_4$, or zero if $y_3 = y_4$. Notice that this adds a new column to the initial tableau that is precisely the negative of the column corresponding to the original coefficients of x_3.

8–15. Show that the values of the two variables suggested in problem 8–14 for replacing an unrestricted variable, x_i, are uniquely determined by the value of x_i.

Hint: You must show that the two y variables cannot both be in the basis at the same time. Thus at least one of the y variables is always zero.

8–16. Solve the following problem where x_1 is unrestricted in sign,

$$x_2 \geq 0, x_3 \geq 0$$

$$5x_1 + 3x_2 + 2x_3 = 2$$

$$6x_1 + 7x_2 + 4x_3 = 12$$

$$10x_1 + 15x_2 + 20x_3 = M, \text{ a maximum.}$$

8–17. Find x_i, $i = 1, 2, 3, 4$, and the minimum value of m where x_2 and x_4 are unrestricted in sign and

$$x_1 \geq 0, \qquad x_3 \geq 0$$

$$5x_1 + 9x_2 + 2x_3 - 2x_4 \leq 19$$

$$8x_1 - 3x_2 \qquad - 4x_4 \geq 9$$

$$7x_1 - 2x_2 + x_3 - x_4 \leq 3$$

$$2x_1 + 5x_2 - x_3 \qquad \geq 3$$

$$x_1 + x_2 - x_3 + x_4 = m.$$

8–18. The Sunnydale Dairy processes 50,000 gallons of raw milk daily. To meet the needs of its regular customers, at least 30,000 gallons will be bottled as fresh milk, at least 6000 gallons will be processed into cheese, and at least 3000 gallons will be used for butter. The manager of Sunnydale will allow at most 4000 additional gallons to be bottled for fresh milk sales to other customers. The processing equipment for cheese and butter can handle no more than a total of 15,000 gallons of raw milk daily. The remainder of the daily supply will be made into powdered milk. However, the manager requires that at least 5000 gallons go to the powdered milk department in order to keep the drying equipment busy. The gross profit from sales figured per gallon of raw milk used in the processing is 50¢ on fresh bottled milk, 10¢ on cheese, 12¢ on butter, and 11¢ on powdered milk. What production schedule subject to these conditions will yield a maximum gross profit to Sunnydale Dairy? What is the shadow price of raw milk that goes into fresh bottled milk? In this case the shadow price represents the increase in profit possible for each additional gallon that the manager allows to go into fresh milk, assuming that it can be sold.

8–19. Ace Advertising Agency has three principal promotion schemes that are carried out by two writers, a designer, a promoter, and a typist. The following table gives the time required in hours for schemes 1, 2, and 3 by each employee and their hours available per month.

	SCHEMES			HOURS AVAILABLE
	1	2	3	
writer A	6	10	9	180
writer B	2	0	8	90
designer	8	6	10	176
promoter	12	5	8	180
typist	4	6	10	160

The three schemes sell for $500, $400, and $1000 respectively. How many customers for each type of scheme are needed per month in order to earn a maximum income? What are the shadow prices per hour of time associated with each employee? Which of the employees would you be willing to hire for overtime to increase company income and what would be a reasonable sum to pay for their overtime?

8–20. Two products from the Stoker manufacturing plant each require the three operations of casting, milling, and machining. The castings for the two products may be bought or made locally. If purchased, casting A costs $4 and casting B costs $8.90. The remaining costs, selling prices, and times involved for products A and B are given in the following table.

	A	B
casting made	$3.50	$7.50
milling	$12.00	$6.00
machining	$2.40	$5.00
selling price	$19.90	$22.45
milling time	1 hr.	½ hr.
machining time	12 min.	25 min.

If made at the local plant, casting A uses 3 lbs. of material C and 1 lb of material D, while casting B uses 5 lbs. of material C and 2 lbs. of material D. There are 8000 lbs. of C and 4000 lbs. of D available for the current two month period. The casting equipment at Stoker's can cast at most 1500 units of A or B during this two months. The milling machinery is available for a total of 1464 hours and the machining equipment for a total of 704 hours during the period. What should the two month production schedule be, and how many of the castings should be bought instead of made? What are the shadow prices of C and D, casting units, hours of milling time, and minutes of machining time? What resources must be increased for an immediate increase in production?

8–21. The Nutrition Problem was published in 1945 by George J. Stigler before the advent of linear programming. At that time the solution was by trial and error because no direct method was available. Solve the following version by linear programming. A consumer desires at least the following quantities of nutrients per day.

proteins	100 gm.
calories	4000
calcium	800 mg.
iron	12 mg.

He decides that his diet will be made up from brown bread, butter, baked beans, cheddar cheese, and spinach. The consumer prices these articles per 100 grams at the local market and then looks up their ingredients in a table of the chemical composition of foods. The results of his research per 100 grams of food are shown in the following table.

	PROTEIN	CALORIES	CALCIUM	IRON	PRICE
	(g)		(mg)	(mg)	($)
brown bread	8.3	246	17.2	2.01	0.26
butter	0.4	793	14.8	0.16	0.74
baked beans	6	93	61.6	2.05	0.21
cheddar cheese	24.9	423	810	0.57	0.60
spinach	5.1	26	595	4	0.19

What should the consumer purchase to satisfy his nutritional requirements and pay the minimum cost per day? What is his minimum cost? Find the shadow prices of protein, calories, calcium, and iron.

8–22. George Dantzig gives the following problem as an example of a linear program for which neither the primal nor the dual has a feasible solution.

$$x_i \geq 0, i = 1, 2$$
$$x_1 - x_2 \geq 2$$
$$-x_1 + x_2 \geq -1$$
$$x_1 - 2x_2 = m, \text{ a minimum.}$$

Solve both the primal and dual programs with the use of artificial variables to show that each is inconsistent. Verify the results graphically.

8–23. Show that the following L.P. has feasible solutions but that the objective is unbounded. Verify that its dual is inconsistent.

$$x_i \geq 0, \ i = 1, 2$$
$$x_1 - x_2 \geq 5$$
$$x_1 - x_2 \geq -5$$
$$-x_1 - x_2 = m, \text{ a minimum.}$$

Check the results graphically. Show that if this problem is run by the Primal-Dual Algorithm, the final tableau has negative entries in both the objective row and the constants column, indicating neither problem has reached an optimum. Such a tableau does not distinguish between unbounded and inconsistent.

8–24. Solve the following L.P. in several ways:

$$x_i \geq 0, i = 1, 2$$
$$3x_1 \geq 3$$
$$-x_1 + x_2 \geq 1$$
$$2x_1 - x_2 \geq -2$$
$$-2x_1 + x_2 \geq 1$$
$$x_1 + 2x_2 \geq -2$$
$$-x_1 + x_2 \geq 1$$
$$3x_1 + 2x_2 \geq -3$$
$$3x_1 + 6x_2 = m, \text{ a minimum.}$$

8–25. Find the solution to the dual of problem 8–24.

8–26. Does the following problem or its dual have feasible solutions?

$$x_i \geq 0, \ i = 1, 2$$

$$-x_1 + \quad x_2 \geq 1$$
$$x_1 - 2x_2 \geq 2$$
$$x_1 + 2x_2 = m, \text{ a minimum.}$$

8–27. Solve the following program and give an alternate optimum.

$$x_i \geq 0, \; i = 1, 2, 3$$
$$2x_1 + \quad x_2 + \quad x_3 \leq 9$$
$$x_1 - \quad 2x_2 + \quad x_3 \geq 2$$
$$4x_2 + \quad x_3 \geq 1$$
$$x_1 + \quad 2x_2 + 2x_3 = m, \text{ a minimum.}$$

8–28. Find the solution to the dual of problem 8–27.

8–29. Solve the following L.P. Is the solution unique?

$$x_i \geq 0, \, i = 1, 2, 3, 4$$
$$x_1 - \quad x_2 - 2x_3 \quad\quad = 0$$
$$2x_1 - 2x_2 - 3x_3 + \quad x_4 = 1$$
$$x_1 - \quad x_2 + \quad x_3 - 4x_4 = M, \text{ a minimum.}$$

8–30. *Approximation Problem.* Suppose $\mathbf{AX} = \mathbf{B}$ is a linear system of m equations in n unknowns that does not have an exact solution. Any vector $\mathbf{X}$ determines an error vector

$$\mathbf{E} = \mathbf{B} - \mathbf{AX}.$$

The problem is to minimize some measure of the error vector $\mathbf{E}$. In analysis, one such measure is the L_p norm, defined as

$$| \mathbf{E} |_p = \left(\sum_{i=1}^{m} | e_i |^p \right)^{1/p},$$

where e_i are the components of vector $\mathbf{E}$. If we let $p = 1$, then the L_1 measure of the error is

$$\sum_{i=1}^{m} | b_i - \sum_{j=1}^{m} a_{ij} x_j |$$

Finding vector **X** such that this error is a minimum is the following linear program. Find vectors **X** and **E** that satisfy

$$\sum_{i=1}^{n} a_{ij} x_j + e_i \geq b_j, \quad i = 1, 2, \ldots, m$$

$$\sum_{j=1}^{n} a_{ij} x_j - e_i \leq b_i, \quad i = 1, 2, \ldots, m$$

such that $\sum_{i=1}^{m} e_i$ is a minimum.

Find vector **X**, vector **E**, and the minimum L_1 error for the system of equations:

$$2x_1 - 3x_2 = 2$$
$$x_1 - 4x_2 = -4$$
$$x_1 + 5x_2 = 15.$$

Plot your solution point and see how close it is to the graphs of the three given lines.

8–31. Another measure of the error vector **E** in problem 8–30 is the L_∞ norm defined as

$$| \mathbf{E} |_\infty = \max | e_i | \quad \text{for} \quad i = 1, 2, \ldots, m.$$

In this case find **X** that minimizes the length of the maximum component of error. The corresponding linear program is: Find vector **X** and number e_o that satisfy

$$\sum_{j=i}^{n} a_{ij} x_j + e_o \geq b_i, \quad i = 1, 2, \ldots, m$$

$$\sum_{j=1}^{n} a_{ij} x_j - e_o \leq b_i, \quad i = 1, 2, \ldots, m$$

such that e_o is a minimum. Number e_o is the length of the maximum component of error. Find vector **X** and e_o for the linear system given in problem 8–30. Are the results different in using these two different norms L_1 and L_∞?

CHAPTER 9

Parametric Programming

9.1 A PARAMETER IN THE OBJECTIVE FUNCTION

One of the difficulties in setting up a linear program is that frequently some of the coefficients in the program are unknown. For example, costs and profits may vary rapidly. Quantities of raw material or inputs may be constantly changing. In general any coefficient in the initial tableau may be known only within some range. One way of solving many programs simultaneously and also trying to analyze the possible variations is to add a parameter to the questionable coefficients. We will discuss two possible cases. One involves a parameter in the objective function and the other involves a parameter in the requirements column. The following problem uses a parameter to handle fluctuations in the objective function.

Example 1. A manufacturer wishes to determine the optimum production schedule for the coming month on three possible products for one of his assembly lines. Last month the three products sold for mark ups of $80, $40, and $50 respectively. Next month the market is uncertain. It is not known how much higher or lower the mark up should be to meet the stiff competition. Still the manufacturer wants to know what the solutions should be depending on possible price variations and what choice of price he has within a given production schedule. Let x_1, x_2, and x_3 respectively be the numbers of the three products to be produced. The production constraints for next month on this assembly line are

$$2x_2 + 3x_3 \leq 800 \tag{1}$$

$$2x_1 + x_2 + 4x_3 \leq 600 \tag{2}$$

$$4x_1 + x_2 + 2x_3 \leq 1000. \tag{3}$$

SOLUTION. To solve the manufacturer's problem, we assume that he wishes to maximize his gross profit determined by the mark up. His objective on last month's run was to maximize

$$80x_1 + 40x_2 + 50x_3. \tag{4}$$

We further assume that each of the three products is subject to the same price fluctuation so that each mark up will be changed by the same amount. By adding a parameter t to the coefficients of (4), we have the objective function for next month:

$$(80 + t)x_1 + (40 + t)x_2 + (50 + t)x_3. \tag{5}$$

To keep track of the coefficients of t under pivoting, we add a row for t at the bottom of our tableau. The initial tableau appears as follows.

	1	2	3	
4	0	2	3	800
5	2	1	4	600
6	(4)	1	2	1000
	−80	−40	−50	0
t	−1	−1	−1	0

The initial tableau is optimal if $t \leq -80$ since the combined objective row will then have all nonnegative entries. This makes economic sense. If the mark up decreases by \$80, it will not be profitable to produce any of the three products. At the borderline case, $t = -80$, we have a possible pivot in the first column. This pivotal column is chosen by a zero entry in the combined objective row, even though such a pivot will not increase the objective value. The result of such a pivot is an alternate optimum. A check of the θ ratios shows this pivot is at (3, 1), giving the second tableau.

	6	2	3	
4	0	2	3	800
5	−.5	(.5)	3	100
1	.25	.25	.5	250
	20	−20	−10	20,000
t	.25	−.75	−.5	250

To see where the second tableau is optimal, it is necessary to examine the inequalities

$$20 + .25t \geq 0$$
$$-20 - .75t \geq 0$$
$$-10 - .5t \geq 0.$$

The solution is

$$t \geq -80$$
$$t \leq -{}^{80}/_{3}$$
$$t \leq -20.$$

All three inequalities are satisfied if t is in the range

$$-80 \leq t \leq -{}^{80}/_{3}. \tag{6}$$

For any value of t in the range of (6), the second tableau is optimal. Choosing the end point $t = -{}^{80}/_{3}$, we find the next pivot in the second column to be at (2, 2). The third tableau is

	6	5	3	
4	(2)	−4	−9	400
2	−1	2	6	200
1	.5	−.5	−1	200
	0	40	110	24000
t	−.5	1.5	4	400

Once again we solve the appropriate inequalities for the range of t that makes the tableau optimal:

$$-.5t \ \geq 0 \quad \text{or} \quad t \leq 0$$
$$40 + 1.5t \ \geq 0 \quad \text{or} \quad t \geq \ ^{-80}/_3$$
$$110 + \quad 4t \ \geq 0 \quad \text{or} \quad t \geq -27.5.$$

The range satisfying all three inequalities for which the third tableau is optimal is

$$-^{80}/_3 \leq t \leq 0. \tag{7}$$

One more iteration is possible by choosing $t = 0$ and finding the pivot in the first column at (1, 1). This pivot produces the fourth tableau.

	4	5	3	
6	.5	−2	−4.5	200
2	.5	0	1	400
1	−.25	.5	1.5	100
	0	40	110	24000
t	.25	.5	1.5	500

The range of t for the fourth tableau to be optimal is all $t \geq 0$.

The four tableaux exhaust all real values for parameter t. The manufacturer may now read off his production schedule from the appropriate tableau for any price fluctuation. For example, if prices go down, he may lower last month's mark up by as much as $26.66 while using the production schedule of tableau three. His solution will be

$$x_1 = 200$$
$$x_2 = 200$$
$$x_3 = \quad 0$$

with a profit of $24,000 + 400t$. A greater reduction in price would force him to use tableau two while an increase over last month's price would result in using tableau four. For any increase in price, $t > 0$, the solution will be

$$x_1 = 100$$
$$x_2 = 400$$
$$x_3 = \quad 0$$

with a profit of $24,000 + 500t$. The manufacturer may set up his production schedule and then make last minute adjustments in price within the appropriate range while being assured of an optimal program. ●

9.2 A PARAMETER IN THE REQUIREMENTS COLUMN

The right-hand column of our initial tableau is called the requirements column. It contains the various inputs into the problem which may very well depend upon a parameter. We will treat such a parameter in a dual way to the method shown in section 9.1. To keep track of the coefficients of the parameter, we add a column to the right side of each tableau. The initial tableau is first optimized and then succeeding pivots are chosen by the Dual Simplex Method. The following example will illustrate the procedure.

Example 2. Find the ranges of the parameter t for which $x_2 + x_3 - x_1$ is a maximum, subject to the constraints

$$x_i \geq 0, i = 1, 2, 3$$

$$2x_1 - 3x_2 + 4x_3 \leq 2 \qquad (1)$$

$$2x_1 + x_2 - 2x_3 \geq t - 2 \qquad (2)$$

$$- x_1 + x_2 + 4x_3 \leq 4 - t \qquad (3)$$

Determine the solution for each of the possible ranges of parameter t.

SOLUTION. The first step in the solution is to multiply constraint (2) by -1 to get a system of type I constraints. Let x_4, x_5, and x_6 be the slack variables. This leads to a typical maximizing program for which the initial tableau is

	1	2	3		t
4	2	-3	4	2	0
5	-2	-1	2	2	-1
6	-1	①	4	4	-1
	1	-1	-1	0	0

The first phase of pivoting is to optimize the tableau by the Simplex Method so that the coefficients in the objective row are all positive. Of course, if there are any artificial variables in the basis, they should be eliminated. In our example a pivot is chosen in the second column at (3, 2). One pivot produces the following optimal tableau.

	1	6	3		t
4	−1	3	16	14	−3
5	(−3)	1	6	6	−2
2	−1	1	4	4	−1
	0	1	3	4	−1

While optimizing the initial tableau the parameter may be ignored except that the t column participates in the pivoting iterations. The optimal tableau is then examined for a range of t that will make it feasible. From the last two columns we have

$$14 - 3t \geq 0$$
$$6 - 2t \geq 0 \tag{4}$$
$$4 - t \geq 0.$$

The three inequalities in (4) are satisfied if $t \leq 3$. For $t \leq 3$ the current tableau is optimal and feasible with solution

$$x_1 = 0$$
$$x_2 = 4 - t$$
$$x_3 = 0$$

where $M = 4 - t$ is maximum. If $t = 3$, basic variable $x_5 = 0$, indicating a degenerate solution. Variable x_5 may be eliminated from the basis by pivoting at (2, 1) according to the Dual Simplex Algorithm. The result is the next tableau.

	5	6	3		t
4	(−.333)	2.667	14	12	−2.333
1	−.333	−.333	−2	−2	.667
2	−.333	.667	2	2	−.333
	0	1	3	4	−1

To find the feasibility range for this optimal tableau, we solve the inequalities

$$12 - \tfrac{7}{3}t \geq 0$$
$$-2 + \tfrac{2}{3}t \geq 0 \tag{5}$$
$$2 - \tfrac{1}{3}t \geq 0.$$

Inequalities (5) are satisfied for t in the range

$$3 \le t \le {}^{36}/_7. \tag{6}$$

For values of t in the range of (6), a solution to the problem is

$x_1 = -2 + {}^2/_3 t$

$x_2 = 2 - {}^1/_3 t$

$x_3 = 0$

$M = 4 - t.$

If $t = {}^{36}/_7$ basic variable $x_4 = 0$, and a pivot chosen in the first row by the Dual Simplex Algorithm occurs at (1, 1), this pivot produces

	4	6	3		t
5	−3	−8	−42	−36	7
1	−1	−3	−16	−14	3
2	−1	−2	−12	−10	2
	0	1	3	4	−1

The last tableau is optimal and feasible for all $t \ge {}^{36}/_7$. The problem solution for t in this range is

$x_1 = -14 + 3t$

$x_2 = -10 + 2t$

$x_3 = 0$

$M = 4 - t.\bullet$

For another example of the use of parameters in linear programming, let us return to example 4, section 8.4, and complete the Sensitivity Analysis suggested there.

Example 3.   In example 4 we found the shadow prices of cotton and dacron to be $43^1/_2$ cents per ounce and 36 cents per ounce respectively. The question was raised as to how much additional cotton could be purchased without changing its shadow price. Likewise we wish to know how much additional dacron can be purchased without changing its shadow price. First we increase only the cotton and let t be the number of ounces of cotton purchased. Then we increase only the dacron and let u be the number of ounces of dacron purchased. The problem is to solve

$$2x_1 + \quad 2x_2 \leq 240 + t \tag{7}$$

$$3x_1 + \quad 2x_2 \leq 260 + u \tag{8}$$

$$3x_1 + \quad x_2 \leq 220 \tag{9}$$

$$1.95x_1 + 1.59x_2 = M \tag{10}$$

for the ranges of t and u subject to the known shadow prices. The two parameters may be added in separate columns to the initial tableau as follows.

	1	2	t	u
3	2	2	240	1	0
4	3	2	260	0	1
5	3	1	220	0	0
	−1.95	−1.59	0	0	0

In three pivots the tableau is optimized, and the final tableau is

	3	4	t	u
5	1.5	−2	60	1.5	−2
2	1.5	−1	100	1.5	−1
1	−1	1	20	−1	1
	.435	.36	198	.435	.36

If $u = 0$, the range of t for which this tableau is feasible can be found from

$$60 + 1.5t \geq 0$$

$$100 + 1.5t \geq 0 \tag{11}$$

$$20 - \quad t \geq 0.$$

The solution to inequalities (11) is

$$-40 \leq t \leq 20. \tag{12}$$

Range (12) shows that 20 additional ounces of cotton can be purchased while the shadow price remains 43½ cents per ounce. The result agrees with our previous answer.

If $t = 0$, the range of u for feasibility is found from

$$60 \; - \; 2u \geq 0$$
$$100 \; - \; u \geq 0 \tag{13}$$
$$20 \; + \; u \geq 0$$

to be

$$- \, 20 \leq u \leq 30. \tag{14}$$

Range (14) shows that 30 additional ounces of dacron can be purchased while the shadow price remains 36 cents per ounce. It should be noted that the columns headed by t and u are identical to those headed by 3 and 4 in the final tableau. It was not necessary to add the columns t and u since each was a unit vector. Recall that in condensing the extended tableau after a pivot, the column unit vector of the variable entering the basis was replaced with the column of coefficients of the variable just removed from the basis. Thus as variables x_1 and x_2 enter the basis, their corresponding columns of unit vectors t and u are replaced with the coefficients of x_3 and x_4 respectively. So it was necessary that column t correspond to x_3 and column u correspond to x_4. This insight grants us a use for the coefficients of a nonbasic variable in the final tableau. They may be used to determine how much a corresponding input can vary while this variable stays out of the basis.   ●

If a variable represents the amount of slack raw material in one of the original problem constraints, then the fact that it is out of the basis in the final tableau means a number of important things. First, the raw material is completely consumed by the optimal solution since its slack equals zero. Second, the shadow price of that raw material is found at the bottom of its column in the objective row. Finally, the remaining coefficients in this column are the multiples of a parameter that, added to the right-hand column and set greater than or equal to 0, will determine the interval over which the corresponding resource may vary in this solution. If this interval is exceeded, the variable in question will be returned to the basis.

Example 4.   One more question of interest from example 4, section 8.4, is how much can the cotton and dacron be simultaneously increased while their shadow prices remain fixed. A limiting factor will be the amount of linen available. To answer this question we add a single parameter to both cotton and dacron in inequalities (7) and (8). The initial tableau is now

	1	2		t
3	2	2	240	1
4	3	2	260	1
5	3	1	220	0
	-1.95	-1.59	0	0

Again after three pivots we reach the optimal tableau

	3	4		t
5	1.5	-2	60	$-.5$
2	1.5	-1	100	$.5$
1	-1	1	20	0
	.435	.36	198	.795

The range of t for which this tableau is feasible is found from

$$60 - .5t \geq 0$$
$$100 + .5t \geq 0$$

to be

$$-200 \leq t \leq 120. \tag{15}$$

Range (15) indicates that cotton and dacron can simultaneously be increased by 120 ounces while their shadow prices remain the same. If we buy an additional 120 ounces of cotton and an additional 120 ounces of dacron, the optimal production schedule is

$$x_1 = 20$$
$$x_2 = 160$$
$$M = 198 + .795(120) = \$293.40.$$

The three slack variables are all zero which means equality holds in the original constraints. In particular all of the slack linen is used up. ●

Our final tableau in parametric form solves a whole range of problems at the same time. Example 4 may be continued by the Dual Simplex Method to find new ranges of the parameter t and the corresponding optimal tableaux. Of course the shadow prices will change as soon as t increases past 120. As an exercise, find the remaining ranges of t, along with the new shadow prices.

PROBLEMS

For the first nine problems, find the optimum value of the objective function for each of the possible ranges of parameter t that both optimizes the corresponding tableau and makes it feasible. All x_i are nonnegative.

9–1. $M = (5 + t)x_1 + (4 + t)x_2 + (8 + t)x_3$ such that

$$5x_1 + 7x_2 + 2x_3 \leq 4000$$
$$2x_1 + 3x_2 + 4x_3 \leq 3000$$
$$4x_1 + x_2 + 2x_3 \leq 5000.$$

9–2. $M = (7 + t)x_1 + (16 - t)x_2$ such that

$$2x_1 + x_2 \leq 3$$
$$x_1 + 4x_2 \leq 4.$$

9–3. $M = (8 + t)x_1 + (6 + t)x_2 + (4 + t)x_3$ such that

$$5x_1 + 4x_2 + 2x_3 \leq 6000$$
$$2x_1 - 1.5x_2 + 5x_3 \leq 8000$$
$$4x_1 + x_2 + 2x_3 \leq 5000.$$

9–4. $M = (5 + t)x_1 + (4 + t)x_2 + (t - 2)x_3$ such that

$$8x_1 + 10x_2 + 5x_3 \leq 1000$$
$$5x_1 + 4x_2 + 12x_3 \leq 710$$
$$4x_1 + 8x_2 + 9x_3 \leq 745$$
$$8x_1 + 5x_2 + 4x_3 \leq 750.$$

In particular what is the maximum and the solution vector if $t = -1$?

9–5. $m = 8x_1 + 3x_2 + 7x_3$ such that

$$x_1 + 2x_2 - 3x_3 \leq 6$$
$$-2x_1 - x_2 + 2x_3 \leq 3 - t.$$

In particular what is the minimum and the solution vector if $t = 5$? $t = 12$?

9–6. $m = 15x_1 + 6x_2 + 4x_3$ such that

$$-x_1 + 2x_2 + 3x_3 \leq 5 + t$$
$$2x_1 + x_2 - x_3 \geq t - 2.$$

From your parametric solution, find the minimum and the solution vector for $t = -6$.

9–7. $m = 5x_1 + 6x_2 + 4x_3$ such that

$$x_1 + 2x_2 + 3x_3 \leq 5$$
$$-2x_1 + x_2 - x_3 \geq t - 2.$$

Is there a solution for $t = 5$?

9–8. $M = x_1 + x_2 + x_3$ such that

$$4x_1 - 5x_2 + x_3 \leq 50 + t$$
$$-5x_1 \qquad - 4x_3 \geq t - 20$$
$$2x_1 + 5x_2 \qquad \leq t + 60.$$

From your parametric solution, find the maximum and the solution vector for $t = -50$.

9–9. $M = 3x_1 + 4x_2 - 2x_3$ such that

$$2x_1 + 3x_2 - x_3 \leq t + 3$$
$$-x_1 - x_2 + 2x_3 \geq 2 - t$$
$$x_1 + x_2 + x_3 = 3.$$

Are there any solutions for $t < -4$? For what value of t is the optimal solution $M = 0$? Does M continue to increase with t?
Hint: Choose the first pivot so as to necessarily eliminate the artificial variable, and then continue by phase II pivoting until the tableau is optimal without regard to t.

9–10. In example 3, section 9.2, find the additional solutions if cotton is increased beyond 20 ounces. What are the shadow prices of dacron and linen for such an increase in cotton?

9–11. In example 3, section 9.2, find the additional solutions if dacron is increased beyond 30 ounces. What are the shadow prices of cotton and linen for such an increase in dacron?

9–12. In example 4, section 9.2, suppose the dacron and cotton are simultaneously increased by more than 120 ounces. Find the possible new solutions along with the new shadow prices. In particular note the increase in the shadow price of linen as the demand for linen increases.

9–13. The Heavy Smelting Company has a plant that produces four alloys by blending together three metals. The resulting alloys, from different combinations of the base metals, achieve various degrees of the desired properties of strength, luster, hardness, and durability. The percentage of each of the base metals A, B, and C that occurs in the four alloys is given in the following table:

	metal		
	A	B	C
alloy 1	60	20	20
2	75	25	0
3	50	40	10
4	80	0	20

There are 1000 tons of metal A, 500 tons of metal B, and 300 tons of metal C available for this month's production. Last month the four alloys sold for a gross profit of $100, $80, $200, and $75 per ton respectively. The selling prices at the end of this month are uncertain. Assume that the four alloys are subject to the same fluctuation in price so that the gross profit on each will be changed by the same amount. We wish to know how Heavy Smelting Company should plan its month's production to maximize gross profits. Determine all of the various possibilities due to possible price changes. In the most likely case of a relatively small change in selling price, what are the shadow prices of metals A, B, and C?

9–14. In problem 9–13, the Heavy Smelting Company has a chance to buy metal B. Suppose that prices remain constant so that this month's gross profit is the same per ton as last month. How many more tons of metal B can be bought without changing the shadow price of B in the optimal solution? What is the maximum profit for each additional ton of B purchased up to this limit?

9–15. In problem 9–13, the Heavy Smelting Company decides to buy more metal A. Suppose that prices remain constant so that this month's gross profit is the same per ton as last month. How many more tons of metal A can be bought without changing the basic variables in the optimal solution? What is the maximum profit for each additional ton of A purchased up to this limit?

9–16. In problem 9–13, is there any price fluctuation for which it would be profitable to buy more metal C?

9–17. Suppose, in problem 9–13, that resources A and B are simultaneously increased. By how many tons can both metals A and B be increased without changing the basic variables in the optimal solution and thus not changing the shadow prices? What is the new profit for each ton increase in both A and B up to this limit? Assume that the profit equation remains the same as last month.

9–18. Continue the analysis of problem 9–17 by increasing both A and B beyond the previous limit. Find each of the possible new limits of increase along with the corresponding new shadow prices. In particular note the demand for metal C as its shadow price increases.

CHAPTER 10

The Transportation Problem

10.1 DEFINITION OF THE PROBLEM

The transportation problem in its direct form is the problem of minimizing the cost of shipping a commodity from a number of origins to various destinations. Suppose there are m origins and n destinations. At the ith origin there are $r_i > 0$ units of the product to be shipped and the number of units required by the jth destination is $s_j > 0$. We are initially given a cost matrix c_{ij}, $i = 1$ to m and $j = 1$ to n, where the entry or *cost coefficient* c_{ij} is the cost of shipping one unit from origin i to destination j. Since the origins correspond to rows and the destinations correspond to columns, the r_i are called *row requirements* and the s_j are called *column requirements*. These quantities may be displayed in tableau form as follows.

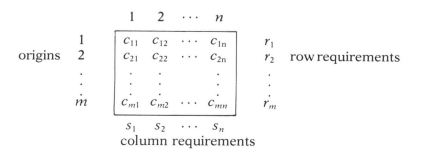

It will be assumed that the total supply equals the total demand, that is

$$\sum_{i=1}^{m} r_i = \sum_{j=1}^{n} s_j. \tag{1}$$

A solution to the problem is a matrix of nonnegative entries X_{ij} equal to the number of units shipped from origin i to destination j

$$\begin{vmatrix} X_{11} & X_{12} & \cdots & X_{1n} \\ X_{21} & X_{22} & \cdots & X_{2n} \\ \cdot & \cdot & & \cdot \\ \cdot & \cdot & & \cdot \\ \cdot & \cdot & & \cdot \\ X_{m1} & X_{m2} & \cdots & X_{mn} \end{vmatrix}$$

such that

$$\sum_{j=1}^{n} X_{ij} = r_i, \text{ for all } i$$

$$\tag{2}$$

$$\sum_{i=1}^{m} X_{ij} = s_j, \text{ for all } j.$$

The first set of equations in (2) says that the sum of each row in the solution matrix is equal to the corresponding row requirement. This means that all of the supply is shipped. The second set of equations in (2) says that the sum of each column in the solution matrix is equal to the corresponding column requirement. This means that demand is satisfied.

Each possible solution matrix has a cost given by

$$\overline{m} = \sum_{i=1}^{m} \sum_{j=1}^{n} c_{ij} X_{ij}. \tag{3}$$

The object of the transportation problem is to find a solution matrix for which the cost $\overline{m}$ is a minimum.

10.2 NORTHWEST CORNER SOLUTIONS

The following simple transportation problem will be used to illustrate the ideas as they are developed.

Example 1. A manufacturer has three factories at different locations that ship a certain product to three distributors around the country. He wishes to minimize his shipping costs. The shipping costs along with the current supply and demand are given in the following tableau.

distributors

		1	2	3		
	1	2	1	5	10	
factories	2	7	4	3	25	inventories
	3	6	2	4	20	

15   18   22
orders

SOLUTION.   The solution set is the 3×3 matrix

X_{11}	X_{12}	X_{13}
X_{21}	X_{22}	X_{23}
X_{31}	X_{32}	X_{33}

$X_{ij} \geq 0$

which must satisfy equations (2) and minimize equation (3). Notice that equation (1) is satisfied, the sum of the row requirements is 55 and the sum of the column requirements is 55. Equations (2) may be written to get $m + n$ equations in mn unknowns.

$$X_{11} + X_{12} + X_{13} = 10$$
$$X_{21} + X_{22} + X_{23} = 25$$
$$X_{31} + X_{32} + X_{33} = 20$$
$$X_{11} + X_{21} + X_{31} = 15 \qquad\qquad (4)$$
$$X_{12} + X_{22} + X_{32} = 18$$
$$X_{13} + X_{23} + X_{33} = 22$$

In this case we have six equations in nine unknowns, but any one equation may be eliminated immediately because the sum of the first three is equal to the sum of the last three. There are at most $m + n - 1 = 5$ independent equations. As we might suspect and will eventually show, the optimal solution has at most $m + n - 1$ variables different from zero. In our example, dropping the last equation and setting $X_{12} = X_{13} = X_{31} = X_{32} = 0$ leads to the triangular system

$$X_{11} \qquad\qquad\qquad\qquad = 10$$
$$X_{22} \qquad\qquad\qquad = 18$$
$$X_{33} \qquad\qquad = 20 \qquad\qquad\qquad (5)$$
$$X_{11} \qquad\quad + X_{21} \qquad = 15$$
$$X_{22} \quad + X_{21} + X_{23} = 25.$$

The system is triangular in the sense that every coefficient above the main diagonal is zero and thus every equation is immediately solvable from the preceding ones. In this case and in general the equations (4) need to be reordered to exhibit the triangular feature in (5). However, the ordering of the equations in the system is immaterial. The important feature is the solvability. We define a system of equations to be *triangular* provided the system can be solved one equation at a time using only the results of the previously solved equations. It is possible to arrange such a system in the form of (5). The solution matrix for equations (5) is

10	0	0
5	18	2
0	0	20

A quick way of arriving at this solution is the Northwest Corner Method. Begin with a blank tableau that has the row requirements to the right and the column requirements underneath.

			10
			25
			20
15	18	22	

Starting at the northwest corner, enter the smaller of the first row or first column requirement. Subtract off this entry from its corresponding row and column requirements.

10			0
			25
			20
5	18	22	

The first row is satisfied. Moving to the second row again enter the smaller of the appropriate requirements and subtract.

10			0
5			20
			20
0	18	22	

The first column is now satisfied. Move to the second column and always enter the smaller of the two requirements.

10			0
5	18		2
			20
0	0	22	

Each time move either right or down according to which requirement remains to be satisfied. The job is complete when all requirements are satisfied.

10			0
5	18	2	0
			20
0	0	20	

10			0
5	18	2	0
		20	0
0	0	0	

It is not necessary to start at the northwest corner. A similar process may be started at any matrix position. We could start at the southwest corner and then use a right-or-up rule to move from cell to cell. An alternate solution to our problem is displayed in the matrix below where the starting position was taken to be X_{32}. A left-or-up rule was used with the convention of associating the left column with the right column as if moving around a circle. Think of the matrix as wrapped around a circular cylinder so that the right and left edges coincide. It may also be necessary to associate the top and bottom edges of the matrix in the same way. Thus when position X_{21} was reached, the next move to the left appears in position X_{23}.

		10	10	0	
13		12	25	12	0
2	18		20	2	0
15	18	22			
13	0	10			
0		0			

The process leads to at most $m + n - 1$ nonzero entries since after the first choice there are $m - 1$ vertical moves and $n - 1$ horizontal moves for a total of $1 + m - 1 + n - 1 = m + n - 1$.

DEFINITION 1. A nonzero X_{ij} is called a *solution variable*. The solution is *feasible* if all solution variables are positive. The solution is *basic* if the solution variables form a triangular system in equations (2).

DEFINITION 2. If a solution contains less than $m + n - 1$ solution variables, then the solution is *degenerate*.

The Northwest Corner Method necessarily leads to a basic feasible solution. Since the solution variables are determined one at a time, the corresponding system of equations is necessarily triangular. The choice of the smaller requirement at each step ensures that each solution variable will be positive so that the result is both feasible and basic.

If we consider starting at every matrix position and tracing out all possible paths, we generate all basic feasible solutions. Thus every basic feasible solution can be obtained by the Northwest Corner process. However, the solution found may be degenerate; this situation will be discussed in the next section.

10.3 DEGENERACY

THEOREM 10-1. *A degenerate basic feasible solution exists if and only if some partial sum of the row requirements equals a partial sum of the column requirements.*

PROOF. Suppose

$$\sum_{i=1}^{p} r_i = \sum_{j=1}^{q} s_j$$

for $p < m$ and $q < n$. Consider the Northwest Corner solution which yields at most $m + n - 1$ solution variables. From our hypothesis, when we arrrive at the pth row and qth column, both a row and a column requirement will be satisfied simultaneously. This forces a diagonal step in the process, moving both a row and a column. Thus there is at least one variable in the solution and the solution is degenerate.

Conversely, if we assume degeneracy and

$$\sum_{i=1}^{p} r_i \neq \sum_{j=1}^{q} s_j$$

for $p < m$ and $q < n$, then the Northwest Corner solution takes the usual $m + n - 1$ steps. This contradicts degeneracy, and therefore equality holds for $p < m$

and $q < n$. If the partial sums do not consist of the first p rows and the first q columns, then the matrix may be reordered until this condition is satisfied. Thus the theorem is true for any partial sum. ■

As an example consider a slight change in example 1, section 10.2.

Example 2.

10			10	0	
4	21		25	21	0
		20	20	0	
14	21	20			
4	0	0			
0					

In this case the sum of the first two row requirements is 35 and the sum of the first two column requirements is 35. When we arrive at the (2, 2) position in the Northwest Corner solution, both the row and column requirements are satisfied. The next move must be a diagonal move to the (3, 3) position. We net only four solution variables instead of the required five for nondegeneracy. The degeneracy may be avoided by perturbing, that is, changing the requirements slightly. One way is by adding a small amount ε to each row requirement and then compensating by adding $m\varepsilon$ to one, say the last, column requirement.

In our case $\varepsilon = .1$ will do.

10.1			10.1
3.9	21	0.2	25.1
		20.1	20.1
14	21	20.3	

The result is five solution variables and the degeneracy is eliminated. ●

In a perturbed problem the correct solution to the original problem may be recovered from the final perturbed tableau by letting ε go to zero. If we define *equivalent transportation problems* to be a pair with the same cost matrix whose corresponding row and column requirements are arbitrarily close to one another, then equivalent problems have the same solution in the limit. The technique of perturbation leads to the following theorem.

THEOREM 10-2. *Every transportation problem with degenerate basic feasible solutions can be replaced with an equivalent problem in which degeneracy is impossible.*

Since degeneracy may be eliminated we need to consider only nondegenerate problems in the rest of the chapter.

10.4 FINDING ADDITIONAL BASIC FEASIBLE SOLUTIONS

Once an initial solution has been found by the Northwest Corner Method, we wish to construct additional basic feasible solutions from the first and determine which is of least cost. The idea will be to introduce a new solution variable to replace one of the current solution variables. This may be done by first introducing a variable ϕ to a vacant cell in the current solution matrix. Then from this cell, trace out what is called the *plus-minus* path. Put a plus in the cell with ϕ and think of ϕ as a positive number added to that cell. To balance the row and column requirements, an amount ϕ must be subtracted from some solution variable in that row and also from some solution variable in that column. Continue to balance the row and column requirements by adding or subtracting ϕ from solution variables until the path returns to the initial cell. To illustrate the process consider the Northwest Corner solution of example 1, section 10.2.

−		+
10		
+	−	
5	18	2
		20

These four vacant cells each determine a unique $(+, -)$ path. If we wish to introduce variable X_{13} into the solution, then the $(+)$ at X_{13} is balanced in the first row by a $(-)$ at X_{11}. Continuing this requires a $(+)$ at X_{21} and a $(-)$ at X_{23}, which balances the original $(+)$, completing the path. The value of ϕ is determined by setting it equal to the smallest X_{ij} with a minus along the $(+, -)$ path. In our case $\phi = 2$. Now adding or subtracting 2 according to the $(+, -)$ path produces the new basic feasible solution.

8		2
7	18	
		20

Note that X_{22} or X_{33} could not be used along this $(+, -)$ path because there would be no way to balance the change. If we introduce variable X_{31} into the Northwest Corner solution, then the $(+, -)$ path begins with a $(+)$ at cell $(3, 1)$ to ensure feasibility. To satisfy the third row requirement, the same amount must be subtracted from 20 in cell $(3, 3)$. Adding the same amount to cell $(2, 3)$ balances the third column. The path is completed by subtracting this amount from cell $(2, 1)$.

10		
—		+
5	18	2
+		—
		20

first B.F.S.

10		
	18	7
5		15

15   18   22

new B.F.S.

For values of ϕ strictly between 0 and 5, we have nonbasic solutions. At $\phi = 5$ we get the new basic solution shown. For $\phi > 5$ the solutions would be infeasible.

If a transportation problem is degenerate it might be impossible to complete a $(+, -)$ path. Consider again example 2, section 10.3.

—		+	
10			10
+	—		
4	21		25
		—	
		20	20

14   21   20

There is no way to balance the second column or the third row. A $(+, -)$ path among the solution variables cannot be completed after starting at the vacant cell (1, 3).

On the other hand we will show that the nondegenerate case determines a unique path. For a nondegenerate transportation problem there can be no complete $(+, -)$ path that goes only through solution variables. If so, then the appropriate value of ϕ would knock out a solution variable leaving the next solution degenerate. Thus, every possible $(+, -)$ path must involve a plus in a vacant cell. If more than one vacant cell is assigned a $(+)$, then the subsequent solutions are nonbasic. Consider ϕ a nonnegative continuous variable assigned to a given vacant cell in our basic feasible solution matrix. Add or subtract ϕ along a $(+, -)$ path from this cell that involves only variables currently in the solution. Allow ϕ to increase continuously from zero. In the resulting solutions some solution variables will increase while others decrease proportionately. Nondegeneracy means that among the decreasing solution variables a unique one will reach zero first. The new feasible solution is therefore unique. If some other $(+,-)$ path through the solution variables were possible from this vacant cell, then a combination from the two would give a complete path among the solution variables contradicting the nondegeneracy. Thus only one path may be found.

We have verified the following theorem.

THEOREM 10-3. *From a nondegenerate transportation problem with a basic feasible solution, a new unique basic feasible solution may be derived starting at each vacant cell of the first solution.*

We certainly don't want to introduce basic solutions at random. The question is which nonsolution variable should be brought into the basis to reduce cost $\overline{m}$ the most? This question will be answered in section 10.6. However, there is a more pressing question at the moment. If a transportation problem has an optimum solution, how do we know that this solution is basic? If the optimum is nonbasic then we are wasting time looking at basic solutions. That a unique optimum must be basic is fundamental to the transportation algorithm.

10.5 THE FUNDAMENTAL TRANSPORTATION THEOREM

THEOREM 10-4. *If a nondegenerate transportation problem has a unique optimum solution, then that solution is a basic feasible solution.*

PROOF.   Start with a nondegenerate transportation problem that has a unique optimum solution. A basic solution for this problem has exactly $m + n - 1$ solution variables. The total possible number of basic solutions is the binomial coefficient

$$\binom{mn}{m+n-1}$$

Ignore those that are infeasible and consider all of the feasible basic solutions to be enumerated with their costs. Let

$$\overline{m} = \sum_{i=1}^{m} \sum_{j=1}^{n} c_{ij} X_{ij}$$

be the one of least cost. We will show that no nonbasic solution can have a smaller cost than $\overline{m}$. Since the optimum solution is unique, only one basic solution can have the cost $\overline{m}$, and all nonbasic solutions will have to cost more. Thus the basic solution with cost $\overline{m}$ will be the optimum solution to the problem.

The demonstration that no nonbasic solution can have a cost smaller than $\overline{m}$ is carried out by induction on the number of solution variables. In the first case we show that this statement is true for $m + n$ solution variables. Introduce a new solution variable X_{ij} into a vacant cell of the basic solution of cost $\overline{m}$. Assume the cost m_1 of this nonbasic solution of $m + n$ variables satisfies

$$m_1 < \overline{m}. \tag{1}$$

Let E_{ij}, called the *entry cost* of X_{ij}, be the cost of adding in one unit along the $(\,+, -\,)$ path from X_{ij}. That is, $\phi = 1$ is added in along the $(\,+, -\,)$ path and E_{ij} stands for the change in cost from cost $\overline{m}$. Thus, for any positive value of ϕ the new cost is

$$m_1 = \overline{m} + \phi E_{ij}. \tag{2}$$

From equation (1), $E_{ij} < 0$. Now let ϕ increase until some solution variable is forced to zero. Then $\phi = \phi_{ij}$ is the value of the incoming variable X_{ij}, which coincides with the value of the unique variable removed from the solution. The cost m_2 of this new basic feasible solution is

$$m_2 = \overline{m} + \phi_{ij} E_{ij} \tag{3}$$

Since E_{ij} is negative, $m_2 < \overline{m}$. This contradicts the fact that $\overline{m}$ was the least cost among basic feasible solutions. The contradiction arose from the assumption in equation (1). Thus, no solution containing $m + n$ variables can have a cost less than $\overline{m}$.

For case two in the induction, we will suppose that no solution with $m + n + k$ solution variables can have a cost less than $\overline{m}$. The object is to show that this statement is true for $m + n + k + 1$ solution variables. Let m_k be the cost of an arbitrary solution with $m + n + k$ solution variables. Then

$$m_k \geq \overline{m}$$

by the induction hypothesis. Let E_{ij} be the entry cost of the $m + n + k + 1$ solution variable and let m_{k+1} be the cost of this solution. Then

$$m_{k+1} = m_k + \phi E_{ij}. \tag{4}$$

If $E_{ij} \geq 0$, then $m_{k+1} \geq m_k \geq \overline{m}$. If $E_{ij} < 0$, then allow ϕ to increase until some solution variable is forced to zero. For this value of ϕ, the solution reduces to the previous case of $m + n + k$ solution variables, and

$$m_k + \phi_{ij} E_{ij} \geq \overline{m} \tag{5}$$

by the induction hypothesis. For any

$$0 < \phi < \phi_{ij}$$

the inequality in (5) is strengthened and combining (4) with (5)

$$m_{k+1} > \overline{m}.$$

Thus no solution with $m + n + k + 1$ solution variables has a smaller cost than $\overline{m}$. The induction gives the truth of this statement for all cases from $m + n$ to mn solution variables and we have completed the proof. ∎

Theorem 10–4 does not answer the questions of existence or uniqueness. The existence question may be answered in the affirmative. In fact no feasible solution can have a cost less than the costs of all basic feasible solutions. Since there are only a finite number of basic feasible solutions, at least one must have the minimum cost. However, the solution may not be unique. If two basic solutions have the same minimum cost, say $m_2 = \overline{m}$, then equation (3),

$$m_2 = \overline{m} + \phi_{ij} E_{ij},$$

implies that the entry cost $E_{ij} = 0$. In this case X_{ij} may be brought into the solution at any value between zero and ϕ_{ij} without changing the cost. For $0 < X_{ij} < \phi_{ij}$ the corresponding solution will be optimal but *not* basic. Let us now return to the question of finding a basic solution of least cost.

10.6 ENTRY COSTS

To find an additional basic feasible solution of less cost, it is necessary to compute the entry costs E_{ij} of the nonsolution variables. The total cost $\overline{m}$ of the new solution may be found from the cost m of the old solution by formula (3) in section 10.5:

$$\overline{m} = m + \phi_{ij} E_{ij}.$$

It is clear from this formula that to reduce cost the E_{ij} must be negative. We will therefore consider only negative entry costs and use the so-called *Rule of Steepest Descent*. Steepest descent means to choose the variable X_{ij} which has the most negative entry cost to enter the next solution. If all entry costs are known, it is then obvious how to proceed.

To compute the E_{ij} easily, we define a new set of variables U_i and V_j by

$$U_i + V_j = c_{ij}, \tag{6}$$

for those *ij* corresponding to a solution variable. Let us return to the Northwest Corner solution of example 1, section 10.2, and write out equation (6).

solution matrix

10			10
5	18	2	25
		20	20
15	18	22	

cost matrix

2	1	5
7	4	3
6	2	4

$$U_1 + V_1 = 2 = c_{11}$$
$$U_2 + V_1 = 7 = c_{21}$$
$$U_2 + V_2 = 4 = c_{22}$$
$$U_2 + V_3 = 3 = c_{23}$$
$$U_3 + V_3 = 4 = c_{33}.$$

We have five equations in six unknowns. Assuming that a transportation problem is nondegenerate, system (6) will lead to $m + n - 1$ equations in $m + n$ unknowns. Since one unknown may be assigned an arbitrary value, let us set $V_1 = 0$. The remaining system is triangular and is immediately solved one equation at a time. The solution is

$$
\begin{aligned}
U_1 &= 2 & V_1 &= 0 \\
U_2 &= 7 & V_2 &= -3 \\
U_3 &= 8 & V_3 &= -4.
\end{aligned}
\qquad (7)
$$

The entry costs for nonsolution variables are listed next in equations (8). Each formula is found by adding or subtracting the costs along the (+, −) path from that variable.

$$
\begin{aligned}
E_{12} &= c_{12} - c_{22} + c_{21} - c_{11} \\
E_{13} &= c_{13} - c_{23} + c_{21} - c_{11} \\
E_{31} &= c_{31} - c_{33} + c_{23} - c_{21} \\
E_{32} &= c_{32} - c_{33} + c_{23} - c_{22}
\end{aligned}
\qquad (8)
$$

Formulas (8) give the change in cost for $\phi = 1$ unit in equation (2). We now substitute equations (6) into equations (8) and simplify.

$$E_{12} = c_{12} - U_2 - V_2 + U_2 + V_1 - U_1 - V_1$$
$$= c_{12} - (U_1 + V_2)$$
$$E_{13} = c_{13} - U_2 - V_3 + U_2 + V_1 - U_1 - V_1$$
$$= c_{13} - (U_1 + V_3) \tag{9}$$
$$E_{31} = c_{31} - U_3 - V_3 + U_2 + V_3 - U_2 - V_1$$
$$= c_{31} - (U_3 + V_1)$$
$$E_{32} = c_{32} - U_3 - V_3 + U_2 + V_3 - U_2 - V_2$$
$$= c_{32} - (U_3 + V_2)$$

In general the Us and Vs with common subscripts will cancel leaving only the U of the row entered and the V of the column entered by E_{ij}. Thus the formula for computing entry costs is

$$E_{ij} = c_{ij} - (U_i + V_j). \tag{10}$$

To write (10) as a matrix equation, we define

$$\mathbf{W}_{ij} = U_i + V_j \quad \text{for} \quad i = 1 \text{ to } m \quad \text{and} \quad j = 1 \text{ to } n. \tag{11}$$

The $\mathbf{W}_{ij}$ matrix is independent of the value arbitrarily assigned to one of the Us or Vs in order to solve system (6). The entry cost for each ij is then the cost matrix minus the $\mathbf{W}_{ij}$ matrix. For ij corresponding to a variable currently in the solution, $W_{ij} = c_{ij}$, and thus $E_{ij} = 0$ for all solution variables.

In the $\mathbf{E}_{ij}$ matrix we need to write only the E_{ij} of nonsolution variables. For our example the result is

$\mathbf{c}_{ij}$		
2	1	5
7	4	3
6	2	4

$-$

$\mathbf{W}_{ij}$		
2	-1	-2
7	4	3
8	5	4

$=$

$\mathbf{E}_{ij}$		
	2	7
-2	-3	

Using the Rule of Steepest Descent, X_{32}, of entry cost -3, should be brought into the solution next to form an improved basic feasible solution.

10.7 THE TRANSPORTATION ALGORITHM

We are now ready to formally state the Transportation Algorithm as an iterative procedure. For a nondegenerate transportation problem it will produce an optimal solution in a finite number of steps. If degeneracy occurs the problem should be perturbed by the method shown in section 10.3 before continuing with the algorithm.

THE TRANSPORTATION ALGORITHM (MINIMUM)

1. Start with an initial basic feasible solution which may be found by the Northwest Corner Method.

2. Construct the $\mathbf{W}_{ij}$ matrix as follows:
   (a) $W_{ij} = c_{ij}$ for ij corresponding to a solution variable.
   (b) Let $V_1 = 0$, and define sets U_i, $i = 1, \ldots, m$, and V_j, $j = 1, \ldots, n$, by $U_i + V_j = c_{ij}$ for those ij corresponding to a solution variable.
   (c) Then the remaining $W_{ij} = U_i + V_j$.

3. Find the entry costs E_{ij} for those variables out of the solution by $E_{ij} = c_{ij} - W_{ij}$. If all $E_{ij} \geq 0$ the solution is optimal.

4. Pick the variable X_{ij} with the most negative entry cost E_{ij} to enter the solution. In the solution matrix trace out the $(+, -)$ path from that variable. Set ϕ equal to the smallest X_{ij} with a $(-)$ in the $(+, -)$ path.

5. Compute a new basic feasible solution by adding or subtracting ϕ along the $(+, -)$ path.

6. Repeat steps 2 through 5 until some solution is optimal, that is, until all entry costs are nonnegative.

7. Note whether $E_{ij} = 0$ for a nonsolution variable in the final tableau. If so, there exist alternate optima that may be found by bringing the variable of zero entry cost into the solution.

If it is desired to find a maximum cost from the algorithm, choose the X_{ij} with the largest positive entry cost to enter the solution. In this case the iterations will terminate when all entry costs are nonpositive. Because there are only a finite number of basic solutions, either the minimizing or the maximizing problem must be terminated in a finite number of steps.

To complete example 1, section 10.2, let us combine all of the calculations into a single matrix. We divide each cell of the solution matrix into three parts as follows.

$$
\begin{array}{|c:c|}
\hline
W_{ij} & E_{ij} \\
\hdashline
\multicolumn{2}{|c|}{X_{ij}} \\
\hline
\end{array}
$$

In addition the values of the solution variables will be circled to make the solution stand out. The V_j will be placed across the top of the tableau and the U_i will be placed in a column to the left. Following the steps of the algorithm in order, we arrive at the initial tableau given below:

cost matrix

$$\begin{array}{ccc} 2 & 1 & 5 \\ 7 & 4 & 3 \\ 6 & 2 & 4 \end{array}$$

solution matrix

	0	−3	−4	
2	2 ⑩	−1 2	−2 7	10
7	7 ⑤	4 3 ⑱−	3 ②+	25
8	8 −2	5 −3 +	4 ⑳−	20
	15	18	22	

The initial tableau contains the Northwest Corner solution whose cost may be found directly by

$$\overline{m} = (\,2\,)10 + (\,7\,)5 + (\,4\,)18 + (\,3\,)2 + (\,4\,)20$$

$$= 213.$$

As before X_{32} at entry cost -3 should come into the solution. The $(\,+,-\,)$ path from X_{32} has been indicated in the initial tableau. We set $\phi = 18$, the smallest solution variable with a $(\,-\,)$ along this path. Adding or subtracting $\phi = 18$ will eliminate X_{22} from the solution and give the next solution.

	0	−6	−4
2	2 ⑩	−4 5	−2 7
7	7 ⑤−	1 3	3 ⑳+
8	8 −2 +	2 ⑱	4 ②−

The cost of the second solution may be found from the first solution

$$\overline{m} = m + \phi_{32}E_{32}$$

$$= 213 + 18(\,-3\,) = 159.$$

The $W_{ij} = c_{ij}$ *for ij* in the solution. These W_{ij} should be filled in first in each tableau. Next set $V_1 = 0$ above the tableau and then $U_1 = 2$, $U_2 = 7$ so that their sum with V_1 agrees with the W values in the first column. The 2 and 7 are placed to the left of the tableau. Next add $U_2 = 7$ with V_3 to get 3 in the W_{23} location. That makes $V_3 = -4$. Now add $V_3 = -4$ with U_3 to get 4 at W_{33}. That forces $U_3 = 8$. Then add $U_3 = 8$ with V_2 to get 2 at W_{32}. So $V_2 = -6$. Now that the Us and Vs are complete, the rest of the $\mathbf{W}$ matrix can be filled in by adding each U with the V above that column. For the variables out of the solution fill in the entry costs, $E_{ij} = c_{ij} - W_{ij}$. That completes the second tableau; all remaining tableaux are computed the same way. All of the computations can be done one at a time, because each of the systems of equations is triangular.

Variable X_{31} should come into the solution next since it is the only nonsolution variable with a negative entry cost. The $(+, -)$ path from X_{31} shows that $\phi_{31} = 2$. Adding or subtracting 2 along the path will eliminate variable X_{33} and give the final tableau.

	0		−4		−4	
2	2 ⑩		−2	3	−2	7
7	7 ③		3	1	3	㉒
6	6 ②		2	⑱	2	2

Since all four entry costs of nonsolution variables are positive, the solution is optimal. The cost of the optimal solution may be found directly or from

$$\overline{m} = 159 + 2(-2) = 155.$$

Thus the minimum possible cost, subject to the original conditions of example 1, section 10.2, is 155. ●

10.8 MULTIPLE SOLUTIONS

The following transportation problem will illustrate the possibility of multiple solutions.

Example 3.

cost matrix   destinations

		1	2	3	
	1	2	1	5	10 row
origins	2	7	3	4	25 requirements
	3	6	5	3	20
		15	22	18	

column
requirements

solution matrix

	0		−4		−6		
2	2 ⑩		−2	3	−4	9	10
7	7 ⑤−		3 ⑳+		1	3	25
9	9 +	−3	5 ②−		3 ⑱		20
	15		22		18		

Variable X_{31} should come into the solution. The (+, −) path determines $\phi = 2$ which eliminates X_{32} from the solution giving

	0		−4		−3		
2	2 ⑩		−2	3	−1	6	
7	7 ③−		3 ㉒		4 +	0	
6	6 ②+		2	3	3 ⑱−		

The cost of this solution from the tableau is

$$\overline{m} = 20 + 21 + 12 + 66 + 54 = 173.$$

Since the entry costs are nonnegative, the solution is optimal and 173 is the minimum cost. However, the zero entry cost for variable X_{23} indicates that this variable may be brought into the solution without changing the cost. Let us introduce X_{23}, choosing $\phi = 3$ which eliminates solution variable X_{21}. The result is an alternate optimum basic solution:

	0		−4		−3	
2	2	⑩	−2	3	−1	6
7	7	0	3	㉒	4	③
6	6	⑤	2	3	3	⑮

The cost of this alternate optimum may be checked to be

$$m = 20 + 30 + 66 + 12 + 45 = 173.$$

The entry costs are the same as in the previous tableau.

Suppose instead of $\phi = 3$ we choose $0 < \phi < 3$ in the next to last tableau. Then the solution obtained is nonbasic but still optimal. For $\phi = 2$ the nonbasic optimal solution is given in the next tableau.

2 ⑩		
7 ①	3 ㉒	4 ②
6 ④		3 ⑯

The cost of this nonbasic solution is also

$$\overline{m} = 20 + 7 + 24 + 66 + 8 + 48 = 173. \quad \bullet$$

We see from this example that the transportation algorithm will necessarily give us an optimal solution but that the optimal solution may not be unique.

It is possible to have a transportation problem that has both degenerate and alternate optima. A degenerate problem might have $m + n - 2$ solution variables in a basic solution instead of the required $m + n - 1$ solution variables for a nondegenerate basic solution. Suppose a problem has two optimal basic solutions of $m + n - 2$ variables. Then nonbasic solutions of $m + n - 1$ solution variables may be formed by choosing ϕ between the two critical values corresponding to the two basic solutions, namely:

$$0 < \phi < \phi_{ij}.$$

The fact that a nonbasic solution might have the right number of variables for a basic solution forces us to define basic in terms of the solution to a triangular system instead of by a count of solution variables. If a nonbasic solution has $m + n - 1$ solution variables as described previously, then it cannot be found by solving a triangular system constructed out of the original equations.

10.9 VARIATIONS

There are a number of possible variations to the transportation problem that widen its applicability. If the context of a problem requires that a certain variable be excluded from the solution, then this variable should be assigned an arbitrarily high cost. By setting the cost high enough, any variable can be forced out of the solution. On the other hand if the problem requires that a certain variable must appear in the solution, then we force this variable into the solution by assigning it a small cost. Provided the cost is small enough, our algorithm will bring the variable into the solution. Then the correct cost of the optimal solution is found by returning this variable to its original cost.

The condition that the sum of the row requirements be equal to the sum of the column requirements may be relaxed. If these sums are not equal we will create a dummy row or a dummy column to take up the slack. A dummy row means that the demand exceeds the supply so that some of the column requirements are not satisfied. A dummy column means that the supply exceeds the demand, so that some items will not be shipped. The cost coefficients of the dummy variables will be taken to be zero so that they will not change the objective function.

It has already been mentioned that the Transportation Algorithm may be used to find a maximum. Instead of changing the choice of incoming variable, an easier way to maximize is to change the sign of all cost coefficients. In this way the same algorithm is used in both cases just as it was in linear programming. The principle involved is that maximizing the negative of the objective function is equivalent to minimizing the objective function. The Transportation Algorithm minimizes the objective function. If this function is replaced with its negative, then the same algorithm will maximize the objective function. Thus a single computer routine will handle both cases. The Transportation Algorithm is quite difficult to program although such programs do exist.[1] However, if the problem is not too large we may solve it by the Simplex Method and use our previously developed automatic routine in BASIC or FORTRAN. The disadvantages of the Simplex Method are that it is quite inefficient compared to the Transportation Algorithm and requires a much larger tableau.

---

[1] G. Bayer, "The Transportation Problem," *Communications of the Association of Computing Machinery* (December 1966): 869, Alg. 293.

10.10 PROGRAMMING THE TRANSPORTATION PROBLEM

In order to use the automatic pivoting routines developed in Chapter 5, we will now solve the transportation problem by the Simplex Method. In comparison, the advantages of the Transportation Algorithm should become apparent.

Example 4. Let us return to example 1, section 10.2, in which the constraining equations were

$$X_{11} + X_{12} + X_{13} = 10$$
$$X_{21} + X_{22} + X_{23} = 25$$
$$X_{31} + X_{32} + X_{33} = 20 \tag{1}$$
$$X_{11} + X_{21} + X_{31} = 15$$
$$X_{12} + X_{22} + X_{32} = 18$$
$$X_{13} + X_{23} + X_{33} = 22$$

$$X_{ij} \geq 0 \quad \text{for} \quad i = 1, 2, 3 \quad \text{and} \quad j = 1, 2, 3. \tag{2}$$

Only $m + n - 1 = 5$ of equations (1) are independent so the last equation in that set will be ignored. The objective function is the cost equation

$$\overline{m} = \sum_{i=1}^{m} \sum_{j=1}^{n} c_{ij} X_{ij}$$

$$= 2X_{11} + 1X_{12} + 5X_{13} + 7X_{21} + 4X_{22} + 3X_{23} + 6X_{31} + 2X_{32} + 4X_{33}. \tag{3}$$

Constraints (1) and (2) along with objective (3) constitute a standard linear program where the objective function is to be minimized. An artificial variable should be introduced for each equality constraint. For convenience let us rename our variables as follows:

$$x_1 = X_{11} \quad x_4 = X_{21} \quad x_7 = X_{31}$$
$$x_2 = X_{12} \quad x_5 = X_{22} \quad x_8 = X_{32}$$
$$x_3 = X_{13} \quad x_6 = X_{23} \quad x_9 = X_{33}$$

The five artificial variables for the first five equations in (1) will be $x_{-10}, x_{-11},$ $x_{-12}, x_{-13},$ and x_{-14}. Each artificial variable is given an arbitrarily large cost, N. Let $\overline{m} = - M$. Then the maximum $\overline{M}$ of the artificial problem is

$$\overline{M} = M - N(x_{-10} + x_{-11} + x_{-12} + x_{-13} + x_{-14})$$
$$\overline{M} + \overline{m} + N(x_{-10} + x_{-11} + x_{-12} + x_{-13} + x_{-14}) = 0.$$

Using our simplified notation, the initial condensed tableau for the Simplex Method is

	1	2	3	4	5	6	7	8	9	
−10	1	1	1	0	0	0	0	0	0	10
−11	0	0	0	1	1	1	0	0	0	25
−12	0	0	0	0	0	0	1	1	1	20
−13	1	0	0	1	0	0	1	0	0	15
−14	0	1	0	0	1	0	0	1	0	18
15	2	1	5	7	4	3	6	2	4	0
16	−2	−2	−1	−2	−2	−1	−2	−2	−1	−88

The objective function has been given a double row with the coefficients of N in the bottom row. Each of the first five rows has been subtracted from the bottom row to get the artificial variables into the basis. The tableau is ready to run. A run on the computer produces the same answer that we arrived at in section 10.7. However, the computer takes 10 pivots compared to the two iterations of the Transportation Algorithm. The Simplex tableau is 7×10 compared to 3×3. The difference becomes more striking as the size of the problem increases. For example in a transportation problem with a 10×10 cost matrix, the Simplex tableau mushrooms to 21×101. A sizable problem may exceed the capacity of your machine. ●

Example 5. As a final technique, let us solve the same problem in example 4 without the use of artificial variables. The artificial variables may be avoided by finding an initial basic feasible solution for the problem variables. Examine the initial tableau above and note that columns 3, 6, and 9 are already unit vectors except for the objective rows. To get two more columns in the same condition, subtract the 5th row from the 2nd row and the 4th row from the 3rd row. The result is the following tableau where columns 3, 5, 6, 7, and 9 are unit vectors except for the objective row.

	1	2	3	4	5	6	7	8	9	
	1	1	1	0	0	0	0	0	0	10
	0	−1	0	1	0	1	0	−1	0	7
	−1	0	0	−1	0	0	0	1	1	5
	1	0	0	1	0	0	1	0	0	15
	0	1	0	0	1	0	0	1	0	18
	2	1	5	7	4	3	6	2	4	0

A basic solution may be completed by subtracting from the objective row 5 times the first row, 3 times the second row, 4 times the third row, 6 times the fourth row, and 4 times the fifth row. This step gives the following extended tableau with a basic feasible solution.

1	2	3	4	5	6	7	8	9	
1	1	1	0	0	0	0	0	0	10
0	−1	0	1	0	1	0	−1	0	7
−1	0	0	−1	0	0	0	1	1	5
1	0	0	1	0	0	1	0	0	15
0	1	0	0	1	0	0	1	0	18
−5	−5	0	2	0	0	0	−3	0	−253

The basic feasible solution, read off from the unit vectors, is $x_3 = 10$, $x_6 = 7$, $x_9 = 5$, $x_7 = 15$, $x_5 = 18$, and $\bar{m} = -M = 253$. The extended tableau may now be condensed by dropping the columns of unit vectors to get an initial tableau for a machine run with our automatic routine.

	1	2	4	8	
3	1	1	0	0	10
6	0	−1	1	−1	7
9	−1	0	−1	1	5
7	1	0	1	0	15
5	0	1	0	1	18
	−5	−5	2	−3	−253

A run of this condensed tableau produces the same familiar solution of cost 155, but the machine requires only three pivots. In addition to greatly reducing the number of pivots, the initial tableau size has been cut down to 6 × 5. The reduction accomplished here might be enough to get the tableau of a large problem down to machine size. ●

Example 6.   A very interesting comparison can be made if, instead of the initial tableau in example 5, we start with the initial condensed tableau corresponding to the Northwest Corner solution. In order to find this tableau let us start again with the initial simplex tableau found from the row and column requirements.

	1	2	3	4	5	6	7	8	9	
1	1	1	0	0	0	0	0	0	10
0	0	0	1	1	1	0	0	0	25
0	0	0	0	0	0	1	1	1	20
1	0	0	1	0	0	1	0	0	15
0	1	0	0	1	0	0	1	0	18
2	1	5	7	4	3	6	2	4	0

Recall that variables x_1, x_4, x_5, x_6, and x_9 are in the Northwest Corner solution. We must produce unit vectors in these five columns. Subtract the first row from the fourth row. Next add the fourth and fifth rows together, and then subtract their sum from the second row. The result shown in the following tableau has unit vectors in columns 1, 4, 5, 6, and 9 with the exception of the objective row.

	1	2	3	4	5	6	7	8	9	
1	1	1	0	0	0	0	0	0	10
0	0	1	0	0	1	−1	−1	0	2
0	0	0	0	0	0	1	1	1	20
0	−1	−1	1	0	0	1	0	0	5
0	1	0	0	1	0	0	1	0	18
2	1	5	7	4	3	6	2	4	0

The Northwest Corner basic feasible solution may be completed by subtracting from the objective row 2 times the first row, 7 times the fourth row, 4 times the fifth row, 3 times the second row, and 4 times the third row. The result of these elementary row operations is the initial extended tableau:

	1	2	3	4	5	6	7	8	9	
1	1	1	0	0	0	0	0	0	10
0	0	1	0	0	1	−1	−1	0	2
0	0	0	0	0	0	1	1	1	20
0	−1	−1	1	0	0	1	0	0	5
0	1	0	0	1	0	0	1	0	18
0	2	7	0	0	0	−2	−3	0	−213

If we drop the columns of unit vectors and place the corresponding variables in the basis, we get this initial condensed tableau:

	2	3	7	8	
1	1	1	0	0	10
6	0	1	−1	−1	2
9	0	0	1	1	20
4	−1	−1	1	0	5
5	1	0	0	(1)	18
	2	7	−2	−3	−213

Compare this initial condensed tableau with the initial solution matrix found by the Transportation Algorithm in section 10.7. Especially note that the shadow prices agree with the entry costs in the Transportation Algorithm. The solution variables have the same values and the cost of this solution, $\overline{m} = -M = 213$, is the same in both cases.

We continue the Simplex Method by pivoting the condensed tableau. The first pivot is at position $(P, Q) = (5, 4)$, and the result is found in the next tableau.

	2	3	7	5	
1	1	1	0	0	10
6	1	1	−1	1	20
9	−1	0	(1)	−1	2
4	−1	−1	1	0	5
8	1	0	0	1	18
	5	7	−2	3	−159

Compare this tableau with the second matrix in the Transportation Algorithm solution. Again everything agrees with the entry costs including the shadow prices. The final pivot is at $(P, Q) = (3, 3)$, and we now look at the final tableau.

	2	3	9	5	
1	1	1	0	0	10
6	0	1	1	0	22
7	−1	0	1	−1	2
4	0	−1	−1	1	3
8	1	0	0	1	18
	3	7	2	1	−155

The shadow prices or entry costs are all positive, so the cost of the minimum solution is 155. Our comparison is complete. The sequence of solutions found by the Simplex Method is identical to the sequence of solutions found by the Transportation Algorithm. The two methods are equivalent, provided only that we start from the same initial basic feasible solution.   •

The simplifying feature of the Transportation Problem is the fact that the coefficients of all of its constraints are zeros or ones. The Transportation Algorithm takes advantage of this feature and thus makes a marked improvement over the general Simplex Method. However our final method in example 6 made similar use of the zeros and ones by finding basic feasible solutions from elementary row operations. A feature of the Transportation Problem is that some set of elementary row operations can always be found that will produce a basic feasible solution. If we are clever enough to find the same initial solution as found by the Northwest Corner Method, then our final method shares all of the advantages of the Transportation Algorithm and gives us the optimal solution in precisely the same number of similar steps. The Simplex Method with its beauty and simplicity is a powerful tool for solving a vast variety of practical problems.

Parametric programming can solve Transportation Problems in the same way illustrated in Chapter 9. A parameter may be added to entry costs or a parameter may be added to the requirements column. This parameter allows many problems to be considered simultaneously.

As long as the row and column requirements are integers, the Transportation Problem always has integer solutions. This very nice feature is guaranteed by the pattern of zeros and ones in the coefficient matrix. Such a matrix is known as totally unimodular. A square matrix is called *unimodular* if its determinant equals one. An m by n matrix, $\mathbf{A}$, is *totally unimodular* if every nonsingular square submatrix is unimodular. All basic solutions to the Transportation Problem,

$$\mathbf{BX} = \mathbf{C} \text{ or } \mathbf{X} = \mathbf{B}^{-1}\mathbf{C}$$

have a coefficient matrix $\mathbf{B}$ that is a nonsingular submatrix of the general coefficient matrix, $\mathbf{A}$. If $\mathbf{A}$ is totally unimodular then $\mathbf{B}$ has determinant 1 and $\mathbf{B}^{-1}$ is an integer matrix. Thus, all solutions $\mathbf{X} = \mathbf{B}^{-1}\mathbf{C}$ are integer whenever $\mathbf{A}$ is totally unimodular and $\mathbf{C}$ is an integer column vector. Most linear programs do not have this property of unimodularity, and special techniques discussed in the next chapter must be applied to get integer solutions.

PROBLEMS

Solve the following transportation problems by the Transportation Algorithm. Perturb if necessary. Indicate those problems that are degenerate and those that have alternate optima. Give an additional optimal solution in cases of alternate optima. In each problem the cost matrix is given along with the row and column requirements.

10–1.

		destinations				
		1	2	3		
	1	45	32	50	100	
origins	2	74	65	81	150	row requirements
	3	28	47	56	75	
		90	120	115		

column requirements

The following is the first published transportation problem.[*]

10–2.

		destinations				
		1	2	3	4	
	1	10	5	6	7	25
origins	2	8	2	7	6	25
	3	9	3	4	8	50
		15	20	30	35	

10–3.

		destinations			
		1	2	3	
	1	7	8	9	150
	2	5	10	2	80
origins	3	4	6	3	170
	4	8	5	12	50
		100	200	150	

10–4.

		destinations				
		1	2	3	4	
	1	2	3	1	4	30
	2	5	7	8	9	80
origins	3	6	9	2	5	50
	4	4	10	6	8	70
		75	60	45	50	

[*] Frank L. Hitchcock, *Journal of Mathematics and Physics*, 20 (1941): 224–230.

10–5. Solve problem 10–4 if variables X_{24} and X_{42} are required to be in the solution by changing their cost coefficients to zero. Does this change increase both of these variables to their maximum possible value?

10–6.

		destinations					
		1	2	3	4	5	
	1	5	3	8	7	5	110
origins	2	7	6	4	5	3	90
	3	8	9	2	4	6	120
		55	40	90	75	60	

10–7. Solve problem 10–6 and determine the minimum cost if the third origin is not permitted to ship to the third destination.

10–8. In problem 10–6, suppose the entries of the cost matrix represent ratings instead of costs. Solve the problem so as to achieve the maximum possible rating.

10–9.

		destinations				
		1	2	3	4	
	1	10	22	16	24	1000
origins	2	12	14	28	36	1500
	3	15	20	35	18	1500
		950	890	900	925	

Note an additional column must be added since supply exceeds demand. How many items will not be shipped from each of the three origins?

10–10. Find the maximum solution to problem 10–9.

10–11.

		destinations				
		1	2	3	4	
	1	10	6	4	8	1000
origins	2	8	12	3	10	1000
	3	7	5	9	2	1000
		800	960	780	850	

In this case, the demand is greater than the supply, so a row must be added. For the optimum solution, what are the shortages at each of the four destinations?

10–12. Find the maximum solution to problem 10–11. Also note the shortages in this case.

The remaining problems in this chapter should be solved by the Simplex Method using the automatic routine developed in Chapter 5. In each case give the solution matrix and the optimum value of the objective function.

10–13. Solve the following transportation problem for minimum cost.

		destinations						
		1	2	3	4	5	6	
	1	2.3	3.2	1.6	5.4	4.8	2.7	5000
origins	2	3.1	5.2	3.6	7.1	2.5	1.9	4000
	3	4.4	6.0	1.5	6.3	3.3	1.7	6000
		2500	2250	2320	980	1350	5600	

10–14. Solve problem 10–13 for the maximum cost.

10–15. Solve the following transportation problem for minimum cost. Is the solution unique?

		destinations					
		1	2	3	4	5	
	1	2.4	4.1	3.3	1.6	5.2	70
origins	2	6.1	3.5	2.9	4.4	1.9	85
	3	2.8	5.3	4.3	3.6	2.6	90
	4	3.2	6.2	1.7	2.8	4.1	65
		42	86	55	67	60	

10–16. Solve problem 10–15 for the maximum cost.

10–17. Solve problem 10–13 without the use of artificial variables by first finding a basic feasible solution as shown in section 10.10.

10–18. Solve problem 10–15 without the use of artificial variables.

10–19. The Flash delivery service has trucks at three locations around town. This morning orders for pick up came in from six different customers. The dispatcher compiles the following mileage chart to show the distance from each truck garage to the various customers:

		customer						
		1	2	3	4	5	6	
	1	10	8.5	12	9.0	3.0	1.5	10
garages	2	3.5	4.0	6.0	5.5	7.0	3.5	8 delivery
	3	15	6.5	11	13	7.5	4.5	11 trucks
		1	2	4	1	3	5	

trucks needed

How should the dispatcher detail his delivery trucks to minimize his cost assuming that the cost is proportional to mileage? What is the minimum required mileage?

10–20. An accident completely ties up one road so that the trucks in problem 10–19 from garage two cannot get through to the third customer. What is the new solution assuming that the dispatcher can reroute his trucks?

10–21. The Seeall Company has two principal factories that produce its color TV sets. The sets are sold through seven distributors around the country. Factory A has a capacity of 5000 sets per month and the capacity of factory B is 7000 sets per month. Next month the requirements of the seven distributors are respectively 1000, 1250, 1850, 900, 2200, 1500, and 2450 TV sets. The TV sets are shipped in lots of 10. The costs per lot of shipping to the distributors are summarized in the following table:

		distributors						
		1	2	3	4	5	6	7
factories	A	$54	62	85	40	105	68	88
	B	$73	58	75	90	73	67	92

What is the required production at each factory next month and what is the shipping schedule to each distributor that will minimize the month's shipping costs?

10–22. A hurricane sweeps inland from the coast leaving a path of destruction with telephone and power lines down in six communities. In four nearby cities the telephone and electric power companies set up emergency crews of linemen to send to the stricken areas. It is essential to get power and communications restored as soon as possible. The linemen travel in crews of two men each. Some crews will be airlifted in by helicopter while others will have to travel by truck. City A has 14 men ready to go in 7 helicopters with pilots. Cities B, C, and D have respectively 18, 14, and 8 men ready to go in their service trucks. The six stricken communities are in need of 6, 10, 8, 12, 6, 12 linemen respectively. The following time chart gives the number of minutes necessary for each crew to reach one of the possible destinations by its mode of travel.

	stricken communities					
	1	2	3	4	5	6
A	18	35	60	24	36	42
B	40	65	38	70	34	80
cities C	45	38	48	55	50	76
D	20	72	25	60	54	94

How should the telephone and electric power companies allocate line crews so as to minimize the total crewminutes of time to reach the disaster areas?

10-23. (*Optimal Assignment Problem*) An executive has four department head positions to be filled. His personnel manager has screened out four men to be hired. Each of the men is given a rating on a scale from 0 to 100 for each of the four jobs. A rating of 100 means that the man is perfectly suited for the job. The ratings are given in the following table:

	job			
	A	B	C	D
1	70	80	75	90
2	75	60	85	78
man 3	82	76	84	88
4	80	68	77	82

Determine which man goes to which job so that the executive achieves the highest total rating for his four departments.

10-24. A machine shop has an order for a number of articles, each of which is composed of four parts that are to be made on four different machines. Four machinists are the employees that run the various machines. From previous work, the shop foreman has composed the following table of costs for each part made by each of the machinists. The costs are given in dollars and cents.

	parts			
	A	B	C	D
1	2.40	4.60	3.55	7.90
2	2.50	4.35	3.25	8.30
machinists 3	2.50	4.25	3.60	8.15
4	2.45	4.10	3.50	7.75

How should the foreman assign his machinists to the jobs so as to minimize the cost of the final product? What is that optimal cost?

10-25. (*Warehouse Problem**) The warehouse problem was first published by Albert S. Cahn in 1948. It is not a transportation problem but rather a

---

* Albert S. Cahn, "The Warehouse Problem," *Bulletin of the American Mathematical Society*, 54 (November 1948): 1073.

general linear programming problem. It is similar to the transportation problem in that the initial tableau is composed of coefficients that are all zeros and ones or minus ones. The initial tableau is totally unimodular, so all tableaux throughout the iterations will be integral. Solve the following example of a warehouse problem.

A certain warehouse will hold 1000 tons of grain. At the beginning of the month there are 600 tons of grain in the warehouse. The manager wishes to buy and sell grain on a weekly basis for the month. His estimated costs C_j per ton and selling prices P_j per ton for the 4 weekly periods $j = 1,2,3,4$, are given in the following table.

		weeks		
	1	2	3	4
C_j	$150	155	125	130
P_j	$170	165	150	140

It is assumed that markets are available so that sales and purchases can be of the desired size at the prevailing P_j and C_j. Sales must be made from the inventory on hand at the beginning of each weekly period. Of course, the amount purchased during any week is limited by the capacity of the warehouse. The problem is to determine the optimal plan of purchasing, storage, and sales so as to maximize the gross profit. Hint: Let $X_j, j = 1, \ldots, 4$ be the number of tons to be purchased in week j. Let $X_j, j = 5, \ldots, 8$ be the number of tons to be sold in week $(j - 4)$. Then the buying constraints due to the capacity of the warehouse are:

$$\sum_{j=1}^{i} X_j - \sum_{j=5}^{i+4} X_j \le 1000 - 600 = 400, \text{ for each } i = 1, \ldots, 4.$$

The selling constraints due to the amount on hand at the beginning of each week are:

$$-\sum_{j=1}^{i-1} X_j + \sum_{j=5}^{i+4} X_j \le 600, \text{ for each } i = 1, \ldots, 4.$$

The nonnegativity constraints are:

$$X_j \ge 0, j = 1, \ldots, 8.$$

The linear function to be maximized is:

$$M = \sum_{j=1}^{4} -C_j X_j + \sum_{j=5}^{8} P_j X_j.$$

In general, if there are *n* periods, the matrix of constraint coefficients is $2n \times 2n$ and is made up of zeros and plus or minus ones.

10-26. Solve problem 10-25 if the costs and selling price per ton are given in the following table.

	weeks			
	1	2	3	4
C_j	$150	140	135	142
P_j	$148	145	150	140

10-27. As general producer of a TV show with four skits, you have responsibility for assigning staff members to do each skit. To produce a skit on time requires four staff members that come from a pool of seven writers, four choreographers, and six musical specialists. The staff members are capable of working on any one of the skits. From experience, you have drawn up the following table rating from 1 to 8 how well each group performs on the four skits, where the lower numbers represent the best performance.

Skits				Staff
1	2	3	4	
2	7	4	2	writers
5	6	3	3	choreographers
3	4	7	5	musical specialists

A staff member works on at most one skit for a given show. You assign a high rating of 8 for any idle staff member. How will you assign your staff to produce the next show and gain the best rating? What staff members are left idle?

10-28. Majestic Motors has assembly lines in Michigan, Maryland, Florida, Texas, and California. A particular deluxe model has a daily output of 100, 90, 120, 80, 90 cars on these assembly lines respectively. There are 12 dealers around the country that carry this special deluxe model. The dealership orders to be filled from today's production are in the following table:

Dealership	1	2	3	4	5	6	7	8	9	10	11	12
order	35	42	28	52	17	33	62	61	43	37	28	42

Decide how many cars each assembly line will ship to each dealership to minimize the total shipping cost. The shipping costs per car to each dealer are given in the next table:

	1	2	3	4	5	6	7	8	9	10	11	12
MI	$ 20	30	100	150	40	70	200	90	60	50	110	140
MD	$ 40	50	90	30	60	100	150	200	250	220	400	350
FL	$100	75	130	40	10	50	80	100	90	300	90	200
TX	$300	340	390	270	100	90	80	110	50	90	75	60
CA	$ 70	80	110	90	55	40	220	310	100	120	80	50

What is the least shipping cost for all of the cars?

CHAPTER 11

Integer Programming

11.1   INTRODUCTION

Many problems require integral solutions because, for example, we cannot buy, sell, or utilize a fractional unit of some product. Unfortunately, the best integer solution is not always the rounded off value of a general solution. We need a method of examining feasible points that have integral coordinates but are not necessarily vertices of the original feasible region. A number of methods have been devised to consider such integral points. These techniques fall into two general categories known as *cutting plane methods* and *enumeration methods*. The cutting plane methods, discussed in sections 11.3 to 11.7, stem primarily from the research of R. E. Gomory beginning in 1958. In the second category of enumeration methods, branch and bound has been most successful. The branch and bound technique, discussed in section 11.7, was devised in the 1960s by several men including A. H. Land, A. G. Doig, J. D. C. Little, and R. J. Dakin. It has been refined since then and is now the base technique of most computer codes used in integer programming.

To introduce the ideas used in cutting planes, we will examine congruences of numbers.

11.2 CONGRUENCE

Let us first define congruence between rational numbers and then see how this idea allows us to represent whole classes of numbers with a single number.

DEFINITION 1. For a pair of rational numbers, a and b, a *is congruent to* b if the difference $a - b$ is a multiple of some integer called the *modulus*.

We will use three bars to stand for *is congruent to*, a notation introduced by the great Carl F. Gauss. Then a is congruent to b modulo n is written $a \equiv b \pmod{n}$ which means $a - b = kn$ where k is an integer. Some examples follow.

$$10 \equiv 1 \pmod 3 \qquad\qquad -\tfrac{1}{2} \equiv \tfrac{5}{2} \pmod 3$$

$$\tfrac{17}{5} \equiv -\tfrac{3}{5} \pmod 4 \qquad\qquad -7 \equiv 1 \pmod 4$$

The congruence relationship with respect to any modulus is reflexive, symmetric, and transitive. These three properties, $a \equiv a$ (reflexive), $a \equiv b$ implies $b \equiv a$ (symmetric), and the two statements $a \equiv b$, $b \equiv c$ imply $a \equiv c$ (transitive) are easily verified. In the case of transitivity note that $a - c = (a - b) + (b - c)$ so that if the latter two are multiples of some modulus so is the former. Any relation that satisfies these three laws is called an *equivalence relation* and numbers equivalent to one another are said to belong to the same *equivalence class*. All numbers congruent to one another belong to the same equivalence class.

Integer programming is based on the modulus 1. Two rational numbers are congruent modulo 1 if and only if their difference is an integer. Given a rational number a we wish to find another rational number b that is congruent to a modulo 1. Of course there are many possibilities. To narrow the field down we will restrict the second number b to lie in the interval $0 \le b < 1$. For convenience let us define a special function to indicate this relationship.

DEFINITION 2. For any rational number a, *function* f is defined by $f(a) = b$, where $a - b$ is an integer and b lies in the interval $0 \le b < 1$.

Some examples of this notation follow:

$$f(\tfrac{1}{7}) = f(\tfrac{43}{7}) = f(-\tfrac{13}{7}) = f(-\tfrac{6}{7}) = \tfrac{1}{7}$$

$$f(-\tfrac{3}{4}) = f(\tfrac{17}{4}) = f(-\tfrac{7}{4}) = f(\tfrac{1}{4}) = \tfrac{1}{4}$$

$$f(5) = f(-9) = f(n) = 0, \text{ for any integer } n.$$

THEOREM 11–1. *For any rational number a, the number b, such that $f(a) = b$, is unique.*

PROOF. Suppose there are two such numbers b for some rational number a; that is, $f(a) = b_1$, $0 \le b_1 < 1$ and $f(a) = b_2$, $0 \le b_2 < 1$. Then by definition

$$a - b_1 = k_1$$
$$a - b_2 = k_2,$$

where k_1 and k_2 are integers. Subtracting these two equations gives $b_2 - b_1 = k_1 - k_2 = k$, where k is an integer. Since both b_1 and b_2 are nonnegative and less than 1, their difference must be strictly between -1 and 1. The only integer in this range is $k = 0$. Therefore, $b_2 = b_1$ and $f(a)$ is unique. ■

Since $f(a)$ determines a unique rational number, we give it a special name.

DEFINITION 3. The number b, determined by $f(a) = b$, is called the *congruent equivalent* to a, that is,

$$a \equiv b \pmod{1}$$

where b is a nonnegative fraction less than 1.

Some further properties of the new notation are given in the theorems below. Congruences will be understood to be (mod 1).

THEOREM 11-2. $f(a + b) \equiv f(a) + f(b)$

PROOF. Let

$$f(a + b) = c, 0 \le c < 1,$$
$$f(a) = d, 0 \le d < 1,$$

and $\quad f(b) = e, 0 \le e < 1.$

We wish to show that $c \equiv d + e$. By the definition of f,

$$(a + b) - c = k_1$$
$$a - d = k_2$$
$$b - e = k_3,$$

where k_1, k_2, k_3 are integers. Subtracting the last two equations from the first equation gives

$$d + e - c = k_1 - k_2 - k_3 = k, \text{ an integer.}$$

Thus $d + e \equiv c$. ■

THEOREM 11-3. *If k is an integer then $f(ka) \equiv kf(a)$.*

PROOF.   Let $f(ka) = c, 0 \leq c < 1$, and $f(a) = d, 0 \leq d < 1$. We wish to show $c \equiv kd$. By the definition of f,

$$ka - c = k_1$$
$$a - d = k_2,$$

where k_1 and k_2 are integers. Multiplying the second equation by k and subtracting from the first gives

$$kd - c = k_1 - kk_2 = k_3, \text{ an integer.}$$

Thus $kd \equiv c$. ∎

Note that if k is not an integer in Theorem 11–3, then it cannot in general be factored out. For example, $f(\frac{1}{2} \cdot \frac{4}{3}) \neq \frac{1}{2}f(\frac{4}{3})$ since $f(\frac{2}{3}) = \frac{2}{3}$ and $\frac{1}{2}f(\frac{4}{3}) = \frac{1}{2} \cdot \frac{1}{3} = \frac{1}{6}$.

THEOREM 11–4.   *If $a \equiv b$ then $f(a) = f(b)$, and conversely, if $f(a) = f(b)$ then $a \equiv b$.*

PROOF.   If $a \equiv b$ then $a - b = k_1$, an integer. Let $f(a) = c, 0 \leq c < 1$, and let $f(b) = d, 0 \leq d < 1$. We must show $c = d$. From the definition of f.

$$a - c = k_2$$
$$b - d = k_3,$$

where k_2 and k_3 are integers. Subtracting the second equation from the first gives

$$a - b + d - c = k_2 - k_3$$
$$k_1 + d - c = k_2 - k_3$$
$$d - c = k_2 - k_3 - k_1 = k,$$

where k is an integer. As in the proof of Theorem 11–1, the only integer in the range of values for $d - c$ is zero. Thus $k = 0$ and $d = c$. Conversely, if $f(a) = f(b)$ then $f(a) - f(b) = 0 \equiv 0$. Using theorem 11–2, theorem 11–3, and transitivity, we have

$$f(a - b) = f(a + (-b))$$
$$\equiv f(a) + f(-b)$$
$$\equiv f(a) - f(b) \equiv 0.$$

To be congruent to zero, $f(a-b)$ must be an integer, but the only integral value of f is zero. Thus $f(a-b)=0$ and $a-b$ must be an integer. This completes the proof that $a \equiv b$. ∎

With the use of theorems 11–1 and 11–4, we now have the rational numbers divided up into equivalence classes (mod 1) so that the congruent equivalent, $f(r)$, for any member r of an equivalence class, is the unique representative of the entire class.

11.3 GOMORY'S FRACTIONAL CUTTING PLANE ALGORITHM

Returning to the problem of finding integral solutions, let us suppose we have an optimal feasible tableau with a nonintegral basic variable. Let x_r be the basic variable that we wish to make integral. Suppose x_r is in the ith row of our optimal tableau, and let $Y_{ij}, j = 1, 2, \ldots, n$, be the tableau entries across that row. The current basic solution involves $x_r = Y_{in}$ where Y_{in} is not integral. If the nonbasic variables are $x_s, x_t, \ldots, x_u$ then the ith row corresponds to the equations

$$x_r + Y_{i1}x_s + Y_{i2}x_t + \cdots + Y_{in-1}x_u = Y_{in} \qquad (1)$$
$$x_r = Y_{in} - (Y_{i1}x_s + Y_{i2}x_t + \cdots + Y_{in-1}x_u).$$

In order to make x_r integral, we want the right-hand side of equation (1) to be congruent to zero, modulo one understood. Of course this condition is currently not satisfied because the nonbasic variables $x_s, x_t, \ldots, x_u$ are all zero. Let us think of these x_j's as unknowns to be brought into the solution by pivoting. Then the condition (2) below will act as an unsatisfied constraint on the problem that will force x_r to be integral when satisfied.

$$Y_{in} - (Y_{i1}x_s + Y_{i2}x_t + \cdots + Y_{in-1}x_u) \equiv 0. \qquad (2)$$

From the definition of congruence, (2) is the same as

$$Y_{in} \equiv Y_{i1}x_s + Y_{i2}x_t + \cdots + Y_{in-1}x_u.$$

By theorem 11–4 we have

$$f(Y_{in}) = f(Y_{i1}x_s + \cdots + Y_{in-1}x_u).$$

Using theorem 11–2 and transitivity

$$f(Y_{in}) \equiv f(Y_{i1}x_s) + \cdots + f(Y_{in-1}x_u). \qquad (3)$$

We eventually want all basic variables to be integral, so we further constrain (3) by considering only integral values of x_j. In this case the x_j may be factored out according to theorem 11–3 giving

$$f(\ Y_{in}\ ) \equiv f(\ Y_{i1}\ )x_s + \cdots + f(\ Y_{in-1}\ )x_u. \tag{4}$$

The right- and left-hand sides of (4) are in the same equivalence class and so differ by an integer. Let

$$f(\ Y_{in}\ ) = \phi, 0 < \phi < 1.$$

Then the right-hand side of (4) is $\phi + k$, for k a nonnegative integer. Thus (4) may be interpreted as a linear constraint with type II inequality as shown in (5).

$$f(\ Y_{i1}\ )x_s + \cdots + f(\ Y_{in-1}\ )x_u \geq f(\ Y_{in}\ ). \tag{5}$$

Furthermore the slack variable for (5), call it x_v, has the integral value $x_v = k$. Subtracting x_v from the left side of (5) and multiplying by (-1) gives the new equation to be added to our optimal tableau.

$$x_v - f(\ Y_{i1}\ )x_s - \cdots - f(\ Y_{in-1}\ )x_u = - f(\ Y_{in}\ ). \tag{6}$$

After adding in equation (6) as a new row, the tableau will be infeasible but remains optimal. Now pivoting can continue by the Dual Simplex Method.

The result of pivoting is to bring new variables into the basis until the new constraint is satisfied. If these new variables come in at integral levels, then the steps in the argument above can be reversed so that congruence (4) implies (3) implies (2). Thus, x_r will be integral. After pivoting terminates, if x_r is not yet integral because new variables came into the basis at nonintegral values, or if nonintegral basic variables remain, the process can be repeated by adding new constraints similar to (5). Eventually both the slack and main variables will be driven to integral values. For a proof of actual convergence of this method in a finite number of steps, refer to Gomory's original papers.[1] The proof assumes that the objective function has a lower as well as an upper bound. If the problem is one of maximizing, then the general solution provides an upper bound for all integer solutions. The existence of a lower bound for integer solutions is guaranteed if the feasible region is bounded. If an integer solution exists it is necessary only to know that the corresponding value of the objective function is greater than some negative constant.

---

[1] Ralph E. Gomory, "Outline of an Algorithm for Integer Solutions to Linear Programs," *Bulletin of the American Mathematical Society*, 64 (September 1958): 275–278; and "An Algorithm for Integer Solutions to Linear Programs," *Princeton IBM Mathematical Research Report* (November 1958).

Each new constraint that is added to the optimal tableau is called a *fractional cutting plane.* A cutting plane removes a part of the feasible region not containing points with integral coordinates. The cuts create new vertices until an optimal integral vertex is found. Since no integral solutions are removed, an optimal solution of the reduced feasible region is an optimal integral solution of the original problem. In starting the algorithm a choice must be made among the rows of the optimal tableau for the one to be used in constructing a fractional cutting plane. There is no sure fire way of making this choice. Experience has shown that it is often quickest to pick the row with the largest fractional part in its right-hand column, that is, for which $f(Y_{in})$ is the greatest. Picking the largest fractional part, $f(Y_{in})$, does not guarantee convergence of the algorithm. Gomory has shown that a cyclic choice, or a random choice among the rows available, will cause the iterations to converge to an integral solution. A summary of the steps in each iteration follows.

GOMORY'S FRACTIONAL CUTTING PLANE ALGORITHM

1. Solve the given problem as a linear program. If there is no finite solution or if the optimal feasible solution is integral, terminate. Otherwise, proceed to step 2.

2. Pick the row from the optimal feasible tableau for which $f(Y_{in})$ is the largest[2] to form the constraint

$$f(Y_{i1})x_s + \cdots + f(Y_{in-1})x_u \geq f(Y_{in}).$$

3. Subtract a slack variable, x_v, and multiply both sides by (-1) to form the equation

$$x_v - f(Y_{i1})x_s - \cdots - f(Y_{in-1})x_u = -f(Y_{in}).$$

4. Place x_v in the basis and add to the optimal tableau the new row of coefficients

$$-f(Y_{i1}), -f(Y_{i2}), \ldots, -f(Y_{in-1}), -f(Y_{in}).$$

5. Continue pivoting by the Dual Simplex Method until the tableau is feasible.

6. If some basic variable is not integral, then repeat steps 2 through 5.

The fractional cutting plane algorithm is very useful for some problems. For example, the author has used it on scheduling problems with several hundred integer variables and gotten optimal integer solutions in a few cuts,

---

[2] In cases that fail to converge, some other choice may work.

generally less than 15. The drawback with this algorithm is that in some other problems, even small problems of five variables, it fails to converge in a reasonable time. Also, if a large number of cuts are required, the coefficients of the cutting constraint become so small that computer round-off error is significant. Because of this round-off error, methods using only integer cuts were devised, and these are presented in sections 11.5 and 11.6. Unfortunately, the integer cuts do not eliminate the convergence problem. Convergence difficulties arise in integer programs that have dual degeneracy, that is, zero coefficients in the objective row. Dual degeneracy causes failures because additional cuts make little or no change in the objective value. Thus, progress towards an optimal objective value is too slow. In general, cutting planes converge rapidly or fail.

The following variation seems to enhance the convergence of the algorithm just presented.

Choose the row for constructing a cut by picking the basic variable in the solution with the largest fractional part, congruent equivalent, between .1 and .9 if available. If no such choice appears, then take the largest part from the remaining choices. Vary both of these choices by using the random number generator on the computer. Pick the largest fractional part in each range with probability ½. If the program does not take the largest part, then it will pick the next largest part with probability ½, etc. Be sure that at least one of the possible choices is taken. This variation incorporates good features from several of the previously mentioned methods of choosing a cut generating row.

11.4 EXAMPLES

For our first example let us solve the following problem by Gomory's Algorithm and graph its solution.

Example 1. Find a pair of nonnegative integers x_1, x_2 that satisfy the constraints

$$- x_1 + 5x_2 \leq 25$$

$$2x_1 + 1x_2 \leq 24.$$

and $10x_2 - x_1$ is maximum. Let $M = 10x_2 - x_1$ or $M + x_1 - 10x_2 = 0$. Using x_3 and x_4 as the slack variables, the initial tableau is easily set up.

	1	2	
3	-1	$\boxed{5}$	25
4	2	1	24
	1	-10	0

The second tableau is found by the Simplex Method, pivoting at position (1, 2).

	1	3	
2	$-1/5$	$1/5$	5
4	$(11/5)$	$-1/5$	18
	-1	2	50

The third and optimal tableau is found by pivoting at (2, 1).

	4	3	
2	$1/11$	$2/11$	$6^8/11$
1	$5/11$	$-1/11$	$8^7/11$
	$5/11$	$21/11$	$58^7/11$

The optimal solution for the original feasible region is

$$x_1 = 8^7/11$$
$$x_2 = 6^8/11$$
$$M = 58^7/11.$$

Unfortunately x_1 and x_2 are not integers, so we apply Gomory's Algorithm. The basic variable with the larger fraction part is x_2. Therefore, we will form our cutting plane constraint from the first row of the optimal tableau as follows:

$$f(1/11)x_4 + f(2/11)x_3 \geq f(6^8/11)$$
$$1/11 x_4 + 2/11 x_3 \geq 8/11.$$

Let x_5 be the new slack variable to be subtracted from the left side of the type II constraint. Then multiply this result by (-1) to form the new equation:

$$x_5 - 1/11 x_4 - 2/11 x_3 = - 8/11.$$

The augmented optimal tableau appears as follows.

	4	3	
2	$1/11$	$2/11$	$6^8/11$
1	$5/11$	$-1/11$	$8^7/11$
5	$\boxed{-1/11}$	$-2/11$	$-8/11$
	$5/11$	$21/11$	$58^7/11$

By the Dual Simplex Algorithm the next pivot is found to be at (3, 1). One pivoting iteration produces a feasible tableau in which all variables are integral.

	5	3	
2	1	0	6
1	5	-1	5
4	-11	2	8
	5	1	55

The best integral solution is

$x_1 = 5$

$x_2 = 6$

$M = 55.$

Since the feasible region has been reduced by the cutting plane, the new maximum is slightly less than the maximum of the original feasible region. The optimal integral vertex (5, 6) cannot be obtained by any rounding off process on the optimal vertex ($8^7/11$, $6^8/11$). As a matter of fact the rounded off answer (9, 7) is not even feasible. Even if we try a nearby feasible point, such as (9, 6) where $M = 51$ or (8, 6) where $M = 52$, the objective function does not reach its maximum for integral coordinates.

A graphical analysis is helpful. We first draw the original feasible region of the given constraints, as shown in Figure 11–1. This region is the quadrilateral with vertices M, N, O, P.

The objective function $-x_1 + 10x_2$ is a family of lines of slope $1/10$. Two members of this family are shown as dashed lines in Figure 11–1. The one through vertex M determines the original maximum of $58^7/11$. If this line of slope $1/10$ is moved into the feasible region keeping its same slope, the first feasible point encountered with integral coordinates is (5, 6) as shown. The actual

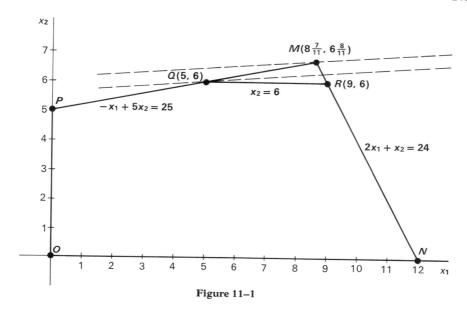

Figure 11–1

Gomory constraint is in the space of variables x_3 and x_4. The cut there determines a new vertex at $x_3 = 0$, $x_4 = 8$. These values of the slack variables in turn determine $x_1 = 5$ and $x_2 = 6$ in the given constraints.

Figure 11–2 shows the Gomory cut as the line segment QR in the plane of variables x_3 and x_4. The remaining constraints in addition to $x_3 \geq 0$ and $x_4 \geq 0$

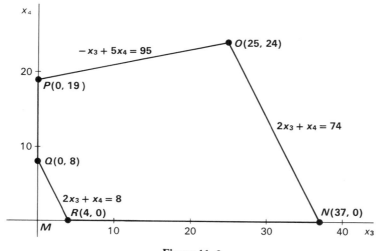

Figure 11–2

may be found from the optimal fractional tableau. After clearing fractions they are

$$2x_3 + \ x_4 \leq 74$$
$$-x_3 + 5x_4 \leq 95.$$

In Figure 11-1 the cut in the plane of x_1 and x_2 that corresponds to the Gomory cut is also labeled QR. It may be determined by substituting the values of x_3 and x_4 from the original equations into the Gomory constraint as follows:

$$2x_3 + x_4 \geq 8$$
$$2(\ 25 + x_1 - 5x_2\ ) + (\ 24 - 2x_1 - x_2\ ) \geq 8$$
$$74 - 11x_2 \geq 8$$
$$x_2 \leq 6.$$

The Gomory cut in Figure 11-2 removes the vertex at the origin and creates new vertices at (0, 8) and (4, 0). In our four dimensional solution space the sequence of vertices corresponding to the basic solutions of our tableaux is in the order

$$O(\ 0, 0, 25, 24\ )$$
$$P(\ 0, 5, 0, 19\ )$$
$$M(\ 8^{7}/_{11}, 6^{8}/_{11}, 0, 0\ )$$
$$Q(\ 5, 6, 0, 8\ ).$$

The cut $x_2 \leq 6$ that we defined in Figure 11-1 did not remove any points with integral coordinates. The integral solution for the reduced region is therefore the same as the integral solution over the original feasible region. In a more complicated problem a number of cuts may be necessary to produce the integral answer. ●

Example 2.   For another example of integer programming let us consider again example 5, section 4.5. The production schedule for minimum cost required the making of a fractional number of bombs. In Chapter 4 we settled for the closest integers as approximate answers. Let us now find the best integer solution to the problem.

SOLUTION.   The original constraints and final tableau are reproduced below where x_1, x_2, and x_3 are the number of bombs to be produced of types A, B, and C respectively.

$$3x_1 + \ x_2 + 6x_3 = 2000 \tag{1}$$

$$2x_1 + 5x_2 + x_3 \geq 1000 \tag{2}$$

$$x_1 + 2x_2 + 4x_3 \leq 3000 \tag{3}$$

$$1600x_1 + 3200x_2 + 2300x_3 = \text{minimum cost.}$$

The slack variables for constraints (2) and (3) were x_4 and x_7 respectively. The final tableau found in section 4.5 was

	2	4	
3	-1.444	.333	111.11
1	3.222	$-.667$	444.44
7	4.556	$-.667$	2111.11
	1366.7	300	$-966,667$

The previous integer solution was

$x_1 = 444$ bombs of type A

$x_2 = 0$ bombs of type B

$x_3 = 111$ bombs of type C.

The solution is quite different after applying Gomory's Algorithm. We attempt to make x_1 integral by adding the new constraint

$$f(3.222)x_2 + f(-.667)x_4 \geq f(444.44)$$

$$.222x_2 + .333x_4 \geq .44.$$

Let x_8 be the new slack variable and then the equation to be added is

$$x_8 - .222x_2 - .333x_4 = -.44.$$

The following optimal tableau is then infeasible.

	2	4	
3	-1.444	.333	111.11
1	3.222	$-.667$	444.44
7	4.556	$-.667$	2111.11
8	$-.222$	$-.333$	$-.44$
	1366.7	300	$-966,667$

By the Dual Simplex Method a pivot is carried out at (4, 2) giving

	2	8	
3	−1.667	1	110.67
1	3.667	−2	445.33
7	5	−2	2112
4	.667	−3	1.33
	1166.67	900	−967,067

The solution is still not integral. Variable x_4 came into the basis at a nonintegral value and so we failed to make x_1 integral. Another cutting plane constraint is needed. This time we work on x_3 to form

$$f(-1.667)x_2 + f(1)x_8 \geq f(110.67)$$

$$.333x_2 + \quad 0 \geq .67.$$

The equation to be added with new slack variable x_9 is

$$x_9 - .333x_2 + 0 = - .67.$$

	2	8	
3	−1.667	1	110.67
1	3.667	−2	445.33
7	5	−2	2112
4	.667	−3	1.33
9	−.333	0	−.67
	1166.67	900	−967,067

After adding in x_9 the above tableau is pivoted at (5, 1) to obtain the final tableau.

	9	8	
3	−5	1	114
1	11	−2	438
7	15	−2	2102
4	2	−3	0
2	−3	0	2
	3500	900	−969,400

The final tableau is completely integral with optimal solution

$x_1 = 438$ bombs of type A

$x_2 = 2$ bombs of type B

$x_3 = 114$ bombs of type C

$m = -M = \$969,400.$

All of the discrepancies in the Chapter 4 answer are now cleared up as the new solution satisfies each constraint precisely with slack $x_4 = 0$ and $x_7 = 2102$. The cost may also be checked in the objective function:

$$\$1600(\,438\,) + \$3200(\,2\,) + \$2300(\,114\,) = \$700,800 + \$6400 + \$262,000$$

$$= \$969,400.$$

Since the cutting planes have reduced the size of the feasible region, our cost for the integral solution is higher than the optimal cost for the original feasible region. However, we have the minimum cost for an integral solution that satisfies all the conditions in the problem. The rounded off solution was actually not in the feasible region. Finally, it is interesting to see that the integral solution even involves a different set of basic variables. Variable x_2 has come into the basis which could not have been predicted by merely looking at the fractional answers. ●

If a problem requires many cuts, the process may be speeded up by adding a number of cutting plane constraints at the same time. Nothing in the finiteness proof restricts the number of constraints that may be added simultaneously. In this case each new constraint is constructed from an appropriate row of the optimal tableau. Pivoting would then continue until the new tableau is feasible.

PROBLEMS

11–1. Use Gomory's Algorithm to solve the following problem for nonnegative integers x_1 and x_2.

$$x_1 + 3x_2 \le 12$$
$$5x_1 + 8x_2 \le 40,$$

where $x_1 + 2x_2 = M$, a maximum. Graph the feasible region for this problem and show a cutting plane through the integral solution.

11–2. Follow the directions to problem 11-1 for solving and graphing the integer solution to the following problem.

$$x_1 + 2x_2 \le 12$$
$$4x_1 + x_2 \le 32,$$

where $2x_1 + 3x_3 = M$ is maximum. Can the integer solution be found by rounding off the optimal general solution?

11-3. Solve for and graph the integer solution to

$$x_1 \geq 0, x_2 \geq 0$$
$$x_1 + 5x_2 \geq 25$$
$$2x_1 - 3x_2 \leq 8,$$

such that $x_1 + 10x_2 = m$, a minimum. On your graph of the feasible region, plot the fractional solutions corresponding to each iteration of Gomory's Algorithm leading up to the integral solution.

11-4. For problem 11-1, graph the Gomory cut in the plane of slack variables x_3 and x_4. Show how this produces a vertex that corresponds to the integer solution for x_1 and x_2.

11-5. Maximize $3x_1 + 6x_2 + 2x_3$ subject to the following constraints where x_i, $i = 1,2,3$, are nonnegative integers.

$$3x_1 + 4x_2 + x_3 \leq 24$$
$$x_1 + 3x_2 + 2x_3 \leq 12$$

11-6. Find nonnegative integers x_i, $i = 1, \ldots, 5$, satisfying the following constraints and yielding the maximum value of the objective function.

x_1					≤ 15
	x_2		$- x_4$		≤ 0
		x_3		$- x_5$	≤ 0
$0.3x_1$			$+ x_4$		$= 9$
$0.18x_1 +$	$0.3x_2$			$+ x_5$	$= 5.4$
$0.06x_1 +$	$0.15x_2 +$	$0.3x_3$		$+$	$7 = M$

List the entire solution vector including the slack variables for the first three constraints. Are the slack variables also integral?

11-7. Does the following problem have a nonnegative integral solution? If so, find it. If not, explain why no integral solution exists and state the general solution.

$$x_1 \qquad\qquad\qquad\qquad \leq 3$$
$$x_2 \quad - \quad x_4 \qquad\qquad \leq 0$$
$$x_3 \quad - \quad x_5 \qquad \leq 0$$
$$0.3x_1 \qquad\qquad + \quad x_4 \qquad = 1.8$$
$$0.18x_1 + \quad 0.3x_2 \qquad\qquad + \quad x_5 \qquad = 1.08$$
$$0.06x_1 + 0.15x_2 + 0.3x_3 \qquad\qquad + \quad 7 = M, \text{ a maximum.}$$

11–8. Find nonnegative integers x_i, $i = 1, \ldots, 5$, such that

$$4x_1 + 3x_2 - 2x_3 + 2x_4 - \quad x_5 \leq 12$$
$$2x_1 + 3x_2 + \quad x_3 + 3x_4 + \quad x_5 \leq 15$$
$$3x_1 + 2x_2 + \quad x_3 + 2x_4 + 5x_5 \leq 20$$
$$2x_1 + 4x_2 + \quad x_3 + 6x_4 + \quad x_5 \leq 25$$
$$x_3 \qquad\qquad \leq 3.$$

where $3x_1 + 2x_2 + 3x_3 + 4x_4 + x_5 = M$ is a maximum.

11–9. Find nonnegative integers x_i, $i = 1, 2, \ldots, 5$, satisfying

$$x_1 \qquad + 2x_3 \qquad + 5x_5 \leq 5$$
$$2x_1 + \quad x_2 + 4x_3 + \quad x_4 + 2x_5 \leq 20$$
$$x_1 + 3x_2 + \quad x_3 + 2x_4 + 3x_5 \leq 33$$
$$3x_1 + 4x_2 \qquad + \quad x_4 \qquad \leq 39.$$

such that $3x_1 + 2x_2 + x_3 + 4x_4 + x_5 = M$, a maximum.

11–10. The Capital Cookie Company makes three types of cookies whose principal ingredients are flour, sugar, chocolate, and butter. The requirements for each type of cookie per pound and resources on hand are listed in the following table in pounds.

	FLOUR	SUGAR	CHOCOLATE	BUTTER
type 1	0.5	0.25	0	0.25
type 2	0.3	0.3	0.2	0.2
type 3	0.25	0.25	0.15	0.35
resources	100	60	50	70

The profit per pound on each type of cookie is 25¢, 29¢, and 24¢ respectively. Type 1 is packed in 5 pound boxes, while type 2 is packed in 2 pound cartons and type 3 is put into 1 pound bags. Determine the best mix of packed cookies and the maximum profit. Which of the four resources would you increase to make the largest increse in profit?

11.5 ALL INTEGER TABLEAUX

If the original tableau has all integer entries, then there is an alternate technique that preserves the integer characteristic in each tableau of all subsequent iterations. This modified simplex algorithm is also due to Gomory.[3] The initial tableau is not first optimized as in the previous method. The idea is to add a cutting plane constraint before each iteration so that the pivot element will be -1. Then a pivot by the Dual Simplex Method on -1 maintains all integers in the new tableau. The initial tableau must be set up as required by the Dual Simplex Method, that is, optimal but not feasible. Then a finite number of iterations with these new constraints leads to an optimal feasible all integer tableau.

The determination of the cutting hyperplane is more difficult than before. It involves the concept of the *greatest integer* in any rational number.

DEFINITION 4. The *greatest integer* in any rational number r is the largest integer less than or equal to r.

The greatest integer in r will be denoted by $[\ r\ ]$. If r is negative, $[\ r\ ]$ is necessarily negative. If $0 \leq r < 1$, then $[\ r\ ] = 0$. Some examples are

$$[\ -\tfrac{1}{3}\ ] = -1$$
$$[\ -7.1\ ] = -8$$
$$[\ \tfrac{9}{10}\ ] = 0$$
$$[\ 5.5\ ] = 5$$
$$[\ -12\ ] = -12.$$

The new constraint will be formed by dividing one of the rows of the current tableau by a rational number $\lambda > 1$, and then taking the greatest integer in each of the resulting numbers. The chosen row must have a negative entry in the constant column Y_{in}. The remaining entries in that row are the coefficients of the nonbasic variables. At least one of these coefficients must be negative or else the problem has no solution. Let the chosen row be the ith row and let the nonbasic variables be $x_s, x_t, \ldots, x_u$. Then the ith row corresponds to constraint

$$Y_{i1}x_s + Y_{i2}x_t + \cdots + Y_{in-1}x_u \leq Y_{in}. \tag{1}$$

---

[3] Ralph E. Gomory, "Research Report RC-189," IBM Research Center, Yorktown Heights, N.Y. (January 1960).

From (1) form the constraint

$$\left[\frac{Y_{i1}}{\lambda}\right] x_s + \left[\frac{Y_{i2}}{\lambda}\right] x_t + \cdots + \left[\frac{Y_{in-1}}{\lambda}\right] x_u \leq \left[\frac{Y_{in}}{\lambda}\right] \quad (2)$$

For $\lambda > 1$ constraint (2) must be satisfied by any integer solution to the original constraint.[4] As long as λ is positive the sense of the inequality is preserved and the sum of the greatest integers cannot be larger than the greatest integer in the sum. By choosing λ large enough, all

$$\left[\frac{Y_{ij}}{\lambda}\right] = -1 \quad \text{for} \quad Y_{ij} < 0.$$

This means that a pivot element of -1 is available in the jth column.

We also wish to keep λ as small as possible while gaining the pivot -1, since a smaller λ produces a greater change in the value of the objective function. This may be seen by substituting these numbers into our formula for the objective value after pivoting in the new row (2).

$$\begin{aligned}
\overline{Y}_{mn} &= Y_{mn} - (Y_{mq} \, Y_{pn} / Y_{pq}) \\
&= Y_{mn} - \left(Y_{mq}\left[\frac{Y_{in}}{\lambda}\right]\right) \Big/ (-1) \\
&= Y_{mn} + Y_{mq} \left[\frac{Y_{in}}{\lambda}\right]
\end{aligned} \quad (3)$$

In a dual program the objective value decreases to its minimum. Because $Y_{mq} \geq 0$ and $Y_{in}/\lambda < 0$, equation (3) shows that the greatest possible decrease in Y_{mn} occurs for small λ.

After λ has been determined, constraint (2) is to be added to the current tableau with a new slack variable in the basis. The new row, call it the pth row, is chosen to be the pivotal row. Then

$$Y_{pj} = \left[\frac{Y_{ij}}{\lambda}\right], \quad j = 1, \ldots, n.$$

The pivotal column q is chosen according to the Dual Simplex Method by the minimum ratio $|\bar{\theta}_q|$ where

$$|\bar{\theta}_q| = \left|\frac{Y_{mq}}{Y_{pq}}\right| = \left|\frac{Y_{mq}}{[Y_{iq}/\lambda]}\right| = \left|\frac{Y_{mq}}{-1}\right| \leq \frac{Y_{mj}}{-[Y_{ij}/\lambda]}$$

for $Y_{ij} < 0$ and $j = 1, \ldots, n - 1$. Thus,

---

[4] T. C. Hu, *Integer Programming and Network Flows*, 1969, pp. 247–248.

$$Y_{mq} \leq \frac{Y_{mj}}{-[\ Y_{ij}/\lambda\ ]} \leq Y_{mj} \qquad\qquad (4)$$

determines the pivotal column to be the one corresponding to the smallest entry in the objective row that is associated with a negative entry on the pivotal row. Let t_j be the largest integer such that

$$Y_{mq} \leq \frac{Y_{mj}}{t_j} \qquad\qquad (5)$$

for $j = 1, \ldots, n-1$ and corresponding $Y_{ij} < 0$.
From (4) and (5) and the definition of t_j, λ must satisfy

$$-[\ Y_{ij}/\lambda\ ] \leq t_j, \qquad \text{for } Y_{ij} < 0. \qquad\qquad (6)$$

Equation (6) says that an arbitrary integer, $-[\ Y_{ij}/\lambda\ ]$ satisfying (4) must be no larger than the greatest possible integer t_j satisfying (5). The smallest λ that will satisfy (6) is

$$\lambda_j = -\ Y_{ij}/t_j \qquad \text{for each possible } j.$$

This is the λ that gives equality in (6). To satisfy (6) for all appropriate j, we pick the largest of the λ_j so the final value of λ is

$$\lambda = \max \lambda_j \qquad \text{for } Y_{ij} < 0, j = 1, \ldots, n-1. \qquad\qquad (7)$$

Notice that our determined value of λ preserves Y_{mq} as the $\bar{\theta}$ ratio of minimum absolute value since the value of the pivot will be -1. For the pivotal column q, $t_q = 1$ and $\lambda_q = -\ Y_{iq}$. Since $\lambda \geq \lambda_q$,

$$0 < \frac{-\ Y_{iq}}{\lambda} \leq \frac{-\ Y_{iq}}{\lambda_q} = 1$$

and thus the pivot $Y_{pq} = [\ Y_{iq}/\lambda\ ] = -1$.

It may happen that $\max \lambda_j = 1$. This means that the generating row already has a pivot of -1 in the qth column. It is preferable to pivot immediately without adding a new constraint. Of course the integer characteristic is preserved. Cutting planes will be constructed only for $\lambda > 1$. The whole procedure is summarized in the following algorithm.

THE ALL INTEGER ALGORITHM

1. Begin with an all integer tableau that is optimal but not feasible.
2. Select a row for generating the cutting plane by a random choice from

those rows with a negative entry in the constant column. For a feasible problem this row, i, must have a negative entry, $Y_{ij} < 0$, for $j < n$.

3. Determine the pivotal column by the minimum entry in the objective row, Y_{mj}, that is associated with a negative Y_{ij}, $j < n$. Call this minimum entry Y_{mq}.

4. Define t_j to be the largest integers such that $Y_{mq} \leq Y_{mj}/t_j$ for $Y_{ij} < 0, j < n$. If $Y_{mq} = 0$, the t_j are undefined. In this case choose $\lambda = |\ Y_{iq}\ |$ and go to step 6.

5. Compute $\lambda_j = -\ Y_{ij}/t_j$, and let $\lambda = \max \lambda_j$.

6. If $\lambda = 1$ then pivot in the ith row at $Y_{iq} = -1$ without adding a constraint. Go to step 2 unless the tableau is feasible.

7. For $\lambda > 1$ construct the constraint in equation (2) by dividing Y_{ij} by λ and taking the greatest integer in each entry for $j = 1, \ldots, n$.

8. Add in the new row with a new slack variable, and pivot in this row at the qth column. The pivot entry will be -1.

9. Repeat steps 2 through 8 until the tableau is feasible or until no negative Y_{ij} can be found with a negative Y_{in}. In the latter case the problem has no integer solution.

The algorithm is explicit except in the choice of a generating row. Gomory has shown in his proof that the process will terminate in a finite number of steps, that either a cyclic or a random choice will work. Our previous choice by the Dual Simplex Method of the row with the largest negative entry in the constant column is not covered under Gomory's proof, although it seems to work in many cases.

To illustrate the whole process, let us again solve example 5, section 4.5, that was handled by Gomory's Fractional Cutting Plane Algorithm in section 11.4. This time we will start with the original constraints replacing the equality constraint with two inequalities. The problem may be stated as follows.

Example 3.   Find nonnegative integers x_1, x_2, x_3 satisfying constraints

$$3x_1 +\ \ x_2 + 6x_3 \leq 2000$$
$$3x_1 +\ \ x_2 + 6x_3 \geq 2000$$
$$2x_1 + 5x_2 +\ \ x_3 \geq 1000$$
$$x_1 + 2x_2 + 4x_3, \leq 3000,$$

where $1600x_1 + 3200x_2 + 2300x_3$ is a minimum.
As usual we let

$$m = -M = 1600x_1 + 3200x_2 + 2300x_3$$

$$M + 1600x_1 + 3200x_2 + 2300x_3 = 0.$$

The type II constraints are multiplied by -1 and the slack variables chosen to be x_4, x_5, x_6, and x_7. The initial tableau is

	1	2	3	
4	3	1	6	2000
5	-3	-1	-6	-2000
6	-2	-5	-1	-1000
7	1	2	4	3000
	1600	3200	2300	0

The initial tableau satisfies step 1 of the All Integer Algorithm. In step 2 we select the second row for generating the cutting hyperplane. All $Y_{2j} < 0$. By step 3 the pivotal column is column 1 and $Y_{mq} = Y_{51} = 1600$. In step 4 we define t_j by

$$1600 \leq 1600/t_1$$

$$1600 \leq 3200/t_2$$

$$1600 \leq 2300/t_3,$$

so that $t_1 = 1$, $t_2 = 2$, $t_3 = 1$. Computing λ_j by step 5 we have

$$\lambda_1 = - Y_{21}/t_1 = {}^3/_1 = 3$$

$$\lambda_2 = - Y_{22}/t_2 = {}^1/_2$$

$$\lambda_3 = - Y_{23}/t_3 = {}^6/_1 = 6$$

$$\lambda = \max \lambda_j = 6.$$

In step 7 we form constraint (2) as follows:

$$[-{}^3/_6]x_1 + [-{}^1/_6]x_2 + [-{}^6/_6]x_3 \leq [-{}^{2000}/_6]$$

$$-1x_1 \qquad -1x_2 \qquad -1x_3 \leq -334.$$

Choose x_8 to be the new slack variable, and then the cutting hyperplane to be added to the initial tableau is

$$x_8 - x_1 - x_2 - x_3 = -334.$$

The new row will be made row 5 so that the objective row is now row 6. The pivot position is at (5, 1) and of course the pivot value is -1. The result of this pivot is to bring x_1 into the basis as shown in the next tableau.

	8	2	3	
4	3	-2	3	998
5	-3	2	-3	-998
6	-2	-3	1	-332
7	1	1	3	2666
1	-1	1	1	334
	1600	1600	700	-534400

We select the second row to generate the new constraint and carry out the steps of our algorithm as follows.

$$Y_{mq} = Y_{63} = 700, \text{ so the pivotal column is column 3.}$$

$$700 \le 1600/t_1, \; t_1 = 2$$

$$700 \le 700/t_3, \; t_3 = 1$$

$$\lambda_1 = {}^3/_2, \; \lambda_3 = {}^3/_1$$

$$\lambda = 3$$

$$[\; -{}^3/_3 \;]x_8 + [\; {}^2/_3 \;]x_2 + [\; -{}^3/_3 \;]x_3 \le [\; -{}^{998}/_3 \;]$$

$$-1x_8 + 0 + (\; -1 \;)x_3 \le -333$$

$$x_9 - x_8 + \quad 0 - x_3 = -333.$$

Slack x_9 is placed in the basis at row 6 and a pivot carried out at (6, 3) to obtain the next tableau.

	8	2	9	
4	0	-2	3	-1
5	0	2	-3	1
6	-3	-3	1	-665
7	-2	1	3	1667
1	-2	1	1	1
3	1	0	-1	333
	900	1600	700	-767500

This time the third row is picked to construct the cutting constraint as follows.

$$Y_{mq} = Y_{71} = 900, \text{ column 1 is pivotal.}$$

$$900 \le 900/t_1, \; t_1 = 1$$

$$900 \le 1600/t_2, \; t_2 = 1$$

$$\lambda_1 = 3, \lambda_2 = 3, \lambda = 3$$

$$[\,-\tfrac{3}{3}\,]x_8 + [\,-\tfrac{3}{3}\,]x_2 + [\,\tfrac{1}{3}\,]x_9 \le [\,-\tfrac{665}{3}\,]$$

$$x_{10} - x_8 - x_2 + 0 = -222.$$

A pivot at (7, 1) in the new row produces the following tableau.

	10	2	9	
4	0	−2	3	−1
5	0	2	−3	1
6	−3	0	1	1
7	−2	3	3	2111
1	−2	3	1	445
3	1	−1	−1	111
8	−1	1	0	222
	900	700	700	−967300

The first row is the only row available for the next cutting constraint. Since only the second column has a negative entry in that row

$$\lambda = -\,Y_{12} = 2$$

$$0 + [\,-\tfrac{2}{2}\,]x_2 + [\,\tfrac{3}{2}\,]x_9 \le [\,-\tfrac{1}{2}\,]$$

$$x_{11} + 0 - x_2 + x_9 = -1.$$

Adding this row and pivoting at (8, 2) we have

	10	11	9	
4	0	−2	3	1
5	0	2	(−1)	−1
6	−3	0	1	1
7	−2	3	6	2108
1	−2	3	4	442
3	1	−1	−2	112
8	−1	1	1	221
2	0	−1	−1	1
	900	700	1400	−968000

The second row has its only negative entry equal to -1, thus we already have a pivot of -1 available. $\lambda = 1$ and by step 6 we pivot at (2, 3) without adding a new constraint. This pivot produces the final tableau.

	10	11	5	
4	0	0	1	0
9	0	-2	-1	1
6	-3	2	1	0
7	-2	15	6	2102
1	-2	11	4	438
3	1	-5	-2	114
8	-1	3	1	220
2	0	-3	-1	2
	900	3500	1400	-969400

The solution agrees precisely with our previous integer solution.

x_1 = 438 bombs of type A

x_2 = 2 bombs of type B

x_3 = 114 bombs of type C

where the minimum cost is \$969,400. Slack variables x_4 and x_5 are necessarily zero since they arose from an equality. Slack variable $x_6 = 0$ corresponds to the previous slack variable x_4 and slack variable $x_7 = 2102$ as before. ●

The same number of pivots, five, was required by both techniques. The All Integer Algorithm has the advantage of avoiding the general solution and keeping all tableaux integral. Each algorithm has advantages on certain problems. However, it is not always possible to tell ahead of time which of the two will result in the fewer iterations on a given problem.

PROBLEMS

The following problems should be done by the All Integer Algorithm. As a check on your work they may also be solved by Gomory's Fractional Cutting Plane Algorithm.

11–11. Find nonnegative integers x_1 and x_2 such that

$$x_1 + x_2 \leq 7 \qquad\qquad x_1 + 6x_2 \geq 6$$

$$2x_1 - x_2 \leq -2 \qquad\qquad 2x_1 - 5x_2 \leq 5,$$

where $x_1 + 14x_2 = m$ is a minimum. Graph the feasible region for this problem and show a cutting plane through the optimal integral point. What is the optimal vertex of the original feasible region?

11-12. Find nonnegative integers x_1 and x_2 such that

$$4x_1 + 7x_2 \leq 28$$
$$2x_1 - x_2 \geq -2$$
$$3x_1 + 10x_2 \geq 15$$
$$2x_1 - 5x_2 \leq 5,$$

where $2x_1 + 24x_2 = m$ is a minimum. Graph the feasible region and show a cutting plane through the optimal integral point. What is the optimal vertex of the original feasible region?

11-13. Work problem 11-12 and answer the same questions if the objective function is changed to $m = 15x_1 + 24x_2$.

11-14. Find nonnegative integers x_i, $i = 1, 2, 3$, satisfying

$$x_1 - 2x_2 + 2x_3 \leq 5$$
$$2x_1 - x_2 - x_3 \geq -13$$
$$x_1 + 2x_2 + 3x_3 \geq 7$$
$$3x_1 + x_2 + 7x_3 \geq 14,$$

such that $2x_1 + 4x_2 + 3x_3$ is a minimum. State the entire solution vector including all slack variables.

11-15. Find nonnegative integers x_i, $i = 1, 2, 3$ satisfying

$$3x_1 + 2x_2 + x_3 \geq 3$$
$$4x_1 + 6x_2 + 5x_3 \geq 6$$
$$x_1 + x_2 + 3x_3 \geq 7,$$

such that $5x_1 + 7x_2 + 10x_3$ is a minimum. State the entire solution vector including all slack variables.

11-16. Find nonnegative integers x_i, $i = 1, 2, 3$, satisfying

$$x_1 + 3x_2 + 2x_3 \geq 5$$
$$2x_1 + x_2 - x_3 \geq 4$$
$$4x_1 - x_2 + x_3 \geq 7,$$

such that $7x_1 + 3x_2 + 2x_3 = m$ is a minimum. Also give the values of all slack variables.

11-17. Find nonnegative integers x_i, $i = 1,2,3,4$, satisfying

$$x_1 + 3x_2 + 2x_3 + 4x_4 \geq 28$$
$$2x_1 + x_2 + 3x_4 \geq 16$$
$$2x_1 + 5x_3 + x_4 \leq 21$$
$$3x_1 + x_2 + 2x_3 + 4x_4 \leq 28,$$

such that $4x_1 + 3x_2 + 2x_3 + x_4 = m$ is a minimum. Give the values of all slack variables.

11-18. Carry out problem 11-17 if the objective function is replaced by the function $m = 4x_1 + 3x_2 + 2x_3 + 5x_4$.

11-19. For the following problem, find the complete solution vector in nonnegative integers.

$$2x_1 + 3x_2 + 5x_3 + 2x_4 + 4x_5 \geq 85$$
$$5x_1 + 2x_2 + 6x_3 + x_4 + 3x_5 \geq 75$$
$$4x_1 + x_2 + 7x_3 + 3x_4 + 5x_5 \geq 99$$
$$3x_1 + 4x_2 + 3x_3 + 5x_4 - 6x_5 \geq 60,$$

where $9x_1 + 5x_2 + 15x_3 + 2x_4 + 8x_5 = m$ is a minimum. Compare the integer solution with the general solution over the uncut feasible region.

11-20. For the following problem find the complete solution vector in nonnegative integers.

$$3x_1 + 5x_2 + 8x_3 + 6x_4 \geq 56$$
$$2x_1 + 7x_2 + 9x_3 - 4x_4 \geq 10$$
$$8x_1 - 3x_2 + 5x_3 + 7x_4 \geq 36$$
$$5x_1 + 6x_2 - 3x_3 + 4x_4 \geq 54$$
$$7x_1 + 4x_2 + 6x_3 + 8x_4 \geq 70$$

where $2x_1 + 4x_2 + 8x_3 + 5x_4 = m$ is a minimum. Compare the integer solution with the general solution over the uncut feasible region.

11-21. Solve problem 11-3 by the All Integer Algorithm.

11-22. Find the integer solution to problem 8-7, section 8.4. Is this solution equivalent to a rounding off of the general solution?

11-23. Find the integer solution to problem 8-11, section 8.4. Can the integer solution be obtained by a rounding off of the general solution?

11-24. Attempt to solve problem 11-1 using the All Integer Algorithm by the following technique. Dualize the problem in order to satisfy the initial conditions. Is the integer solution to the dual equivalent to the integer solution to the primal problem? Notice that the von Neumann Minimax Theorem only holds in the continuous case and not in the discrete case of integer solutions.

11-25. Dualize problem 11-2 and find the minimum by the All Integer Algorithm. Is this minimum the same as the maximum to the primal problem found by Gomory's Fractional Cutting Plane Algorithm?

11-26. The Boxcar Flying Service Company has a rush order to deliver supplies and parts to a neighboring country. They are packed in crates of two sizes. The volumes and weights of each size are given in the following table.

		cu. ft.	lbs.
sizes	1	60	400
	2	50	420

Two planes are available for today's run, Sue and Flo. Today's initial delivery must consist of at least 22 size 1 crates and at least 30 size 2 crates. Sue has a maximum load limit of 10,000 pounds of cargo and the limit for Flo is 12,000 pounds of cargo. The efficiency expert of Boxcar requires that Sue carry at least 1300 cubic feet of cargo on each flight while Flo must carry at least 1520 cubic feet of cargo on each flight. The total cost of handling and shipping these crates is $3 per pound on Sue and $3.50 per pound on Flo. What is the minimum cost of today's delivery and how should Boxcar load the crates on its two planes so as to achieve the minimum cost? Compare your integer solution to the solution found without using any cutting plane constraints.

11.6 EXTENSIONS OF THE ALL INTEGER ALGORITHM

To make the All Integer Algorithm apply to primal programs where the initial tableau is feasible but not optimal, let us introduce an upper bound constraint. It is usually easy from the given constraints to determine an upper bound on the sum of the main variables. Of course one way to find a bound is to solve for the general solution and then use the next integer larger than the actual sum of these variables. The upper bound constraint is

$$x_1 + x_2 + \cdots + x_{n-1} \leq B \qquad\qquad (1)$$

where B is the predetermined integer bound. If constraint (1) is added into the nonoptimal tableau, and then a pivot is performed in this row at the column with the most negative entry in the objective row, then the new tableau will be optimal but probably not feasible. Now the All Integer Algorithm may be applied until the tableau is feasible. All tableaux remain integral if the initial tableau is integral.

Example 4. Consider example 1, section 11.4, that was solved by Gomory's Fractional Cutting Plane Algorithm. The general solution was $(8^{7}/_{11}, 6^{8}/_{11})$ so an obvious upper bound on the sum is $B = 16$. Let the upper bound constraint be

$$x_1 + x_2 \le 16.$$

If the new slack variable is x_5, then the initial tableau is as follows

	1	2	
3	−1	5	25
4	2	1	24
5	1	①	16
	1	−10	0

A pivot at (3, 2) produces the following optimal infeasible tableau.

	1	5	
3	−6	−5	−55
4	1	−1	8
2	1	1	16
	11	10	160

Using the All Integer Algorithm on row one, we find $\lambda = 6$, and the cutting plane is

$$-x_1 - x_5 \le -10.$$

By adding in the new row with slack x_6 and pivoting at (4, 2), we have the next tableau.

	1	6	
3	(−1)	−5	−5
4	2	−1	18
2	0	1	6
5	1	−1	10
	1	10	60

This tableau has a natural pivot of − 1 at (1, 1). Carrying out the pivoting gives the final tableau.

	3	6	
1	−1	5	5
4	2	−11	8
2	0	1	6
5	1	−6	5
	1	5	55

The integer solution is the same as that found previously.

$M = 55$ at (5, 6, 0, 8). ●

The purpose of the All Integer Algorithm is to maintain the integer characteristic in all tableaux. However, there is nothing in the construction of the algorithm or in the proof of its finiteness that requires the constraint coefficients to be integers. In other words the algorithm may be applied to a tableau with real coefficients provided the objective row has all integer entries. Naturally the cutting plane has integer coefficients and the pivot will still be − 1. Thus all variables coming into the basis will come in at integral levels. When the tableau is feasible the main variables will be integral in the solution. Slack variables may remain nonintegral.

The choice of a generating row for a cutting plane is important. Some cuts are stronger than others in the sense that fewer iterations are required to finish the job. In general if a natural pivot of −1 can be found in the tableau, it will be stronger than an arbitrary cut. This follows since a cut is computed from $\lambda > 1$, while $\lambda = 1$ produces a greater change in the objective value. As has been noted in step 6 of the All Integer Algorithm, a pivot should be taken in this case without adding a constraint. It is wise to look for a natural pivot of −1 while searching for a generating row. A good example of the improvement possible occurs in problem 26, section 11.5. This problem may be completed in five iterations if the natural pivots are taken. If a cutting plane is added at every step, the problem mushrooms to over a dozen iterations.

A final observation is helpful in controlling the tableau size. Both the fractional and the all integer algorithms increase the tableau size with each cutting plane. After a number of iterations some of the cutting plane slack variables that were removed from the basis will be returned to the basis. A variable coming into the basis by pivoting necessarily comes in with a nonnegative value. The corresponding constraint for this reentering slack variable is thus satisfied by the current solution. This row may be immediately dropped from the tableau since it is redundant. The advantage is that after a certain point the tableau size will not increase as one row is dropped after each iteration. If a redundant row is not dropped, then at some later iteration it may again have a negative value in its last column. When this occurs the row may be used as a pivotal row or a cut generating row. However, there is little advantage to keeping the extra row. Further extensions will be considered in the problems.

PROBLEMS

11–27. Use an upper bound constraint along with the All Integer Algorithm to solve the following problem for nonnegative integers x_1 and x_2.

$$x_1 + 3x_2 \leq 12$$
$$5x_1 + 8x_2 \leq 40$$
$$x_1 + 2x_2 = M, \text{ a maximum.}$$

11–28. Follow the directions to problem 11–27 for finding the integer solution to

$$x_1 + 2x_2 \leq 12$$
$$4x_1 + x_2 \leq 32$$
$$2x_1 + 3x_2 = M, \text{ a maximum.}$$

Compare your result to the previous solution found by Gomory's Fractional Cutting Plane Algorithm.

11–29. Find nonnegative integers x_1 and x_2 satisfying

$$3x_1 - x_2 \geq -3$$
$$x_1 + 18x_2 \leq 109$$
$$23x_1 + 6x_2 \leq 161$$
$$4x_1 + 15x_2 \geq 20,$$

such that $x_1 + 3x_2$ is maximum. Compare your solution with the nonintegral solution.

11-30. Find nonnegative integers x_1 and x_2 such that $x_1 + 9x_2$ is maximum, subject to the same constraints given in problem 11-29. Compare your solution with the nonintegral solution.

11-31. Find nonnegative integers x_1 and x_2 such that $7x_1 + 15x_2$ is minimum, subject to the same constraints given in problem 11-29. Compare your solution with the nonintegral solution.

11-32. Find nonnegative integers x_1 and x_2 such that $4x_1 - x_2$ is minimum, subject to the same constraints given in problem 11-29. Compare your solution to the general solution over the uncut feasible region.

11-33. Solve problem 11-6, section 11.4, by the All Integer Algorithm. Hint: Convert all coefficients into integers.

11-34. An intuitive Primal All Integer Algorithm may be developed as follows:

    a. Start with a tableau that is feasible but not optimal and has all integer entries.

    b. Choose the pivotal column q by the most negative number in the objective row.

    c. Choose the source row i by the minimum θ ratio for all positive entries in the pivotal column q.

    d. Let $\lambda = Y_{iq}$ for the chosen ith row and qth column. If $\lambda = 1$ then pivot on Y_{iq} and go to step g.

    e. If $\lambda > 1$ construct a cutting constraint by dividing the source row i by λ and then taking the greatest integer in each coefficient.

    f. Add the cutting constraint with a new slack variable to the tableau and pivot in this row at the qth column.

    g. Repeat steps b through f until the tableau is optimal or unbounded.

Try this algorithm on example 4, section 11.6.

$$x_1 \geq 0, x_2 \geq 0$$
$$-x_1 + 5x_2 \leq 25$$
$$2x_1 + x_2 \leq 24,$$

where $-x_1 + 10x_2 = M$ is a maximum. Note that after redundant rows are dropped the solution is the same as before but requires more iterations. Unfortunately this algorithm does not admit a finiteness proof without further refinements.[6] However, it will work if the problem does not run into a cycle due to degeneracy.

---

[6] F. Glover, "A New Foundation for a Simplified Primal Integer Programming Algorithm," *Journal of the Operations Research Society of America*, 16 (4) (July–August 1968): 727–740.

11-35. Find the complete solution vector to the following problem where $x_1, x_2,$ and x_3 are nonnegative integers. Assume the given coefficients are real numbers not necessarily rational.

$$3.1416x_1 + 2.718x_2 + 1.414x_3 \geq 12$$
$$1.732x_1 + 2.236x_2 + 2.718x_3 \geq 14$$
$$2.111x_1 + 3.1416x_2 + 2.54x_3 \geq 15,$$

where $x_1 + x_2 + x_3 = m$ is a minimum.

11-36. Solve problem 11-35 where the objective function is changed to $x_1 + 2x_2 + 3x_3 = m$.

11-37. It is possible to put all of the variables into the basis of the initial tableau. The nonbasic variables may be added to the basis by throwing in the identities $x_q - x_q = 0$ for each appropriate q. Try this on the example in problem 11-34 and show the initial tableau is

	1	2	
1	-1	0	0
2	0	-1	0
3	-1	5	25
4	2	1	24
	1	-10	0

Carry out the Primal All Integer Algorithm by adding in each cutting constraint as row 5 and then dropping row 5 immediately after the pivot. The advantage is that the tableau remains the same size and the final tableau has x_i, $i = 1$ to 4, in the first four rows respectively.

11.7 THE BRANCH AND BOUND METHOD

One of the popular and powerful techniques of integer programming is known as the *branch and bound method*. It is the foundation of most of the commercial computer codes. The term *branch and bound* was first used by J. D. C. Little in an algorithm he provided in 1963 for solving the traveling salesman problem. Some earlier work on these ideas was done by A. H. Land and A. G. Doig in 1960. The method has been expanded by R. J. Dakin in 1965 and many others since then to make the bounding more efficient.

The idea behind this method is to subdivide the feasible region into a number of subregions that together contain all the feasible points with integer coordinates, called *grid points*. A subdivision is created by removing a strip between adjacent grid points. The removal process is continued systematically until the remaining subregions have a grid point for the optimal vertex of that subregion. If the initial feasible region is bounded, that is, the sum of all the variables for a feasible solution is finite, then there exists only a finite number of these strips to be removed. Thus, the process will terminate in a finite number of steps. Even if the feasible region is unbounded, the process is finite. The integer program is unbounded or has no solution if the corresponding linear program (L. P.) is unbounded. As long as the L. P. is feasible and finite, the subregions created will require only a finite number of subdivisions. Since no grid points were removed, the best integer solution found among the finite number of subregions is the global integer solution over the original feasible region.

The principal difficulty with this subdividing or branching technique is that an unreasonably large number of subregions may need to be investigated. Some method must be devised to eliminate those subregions that are unpromising and concentrate on relatively few subregions that may contain an optimal integer solution to the original problem.

This is where the bounding comes in. If a lower bound on the objective maximum for integer solutions can be determined, then all of the subregions known to have an optimal objective value less than this lower bound are eliminated. The eliminated subcases are said to be "fathomed." At first a rough estimate may be used for this lower bound. Clearly the tighter the lower bound on the objective maximum, the more subcases that can be eliminated from the search. The significant progress made with this method since its inception has been in improving the bounding.

In order to see how the branching is carried out, consider the following example:

Example 5.

$$-2x_1 + 5x_2 \leq 20$$
$$21x_1 + 6x_2 \leq 85$$

where $31x_1 + 9x_2$ is to be maximized and x_1, x_2 are nonnegative integers.

The solution is shown in Figure 11–3 along with a "tree" of the various branches. The first step is to solve the integer program as if it were a linear program to see what the solution is in real numbers over the original feasible region determined by the given constraints. Of course, if the L. P. solution turns out to be integer, the problem is solved. Otherwise, pick a variable from the L. P. solution that has a nonintegral value as the branching variable. This variable determines two integer branches, one by considering the next lower integer and the other by taking the next higher integer.

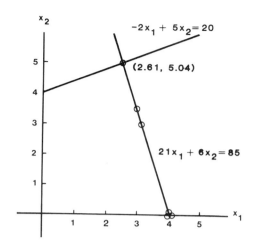

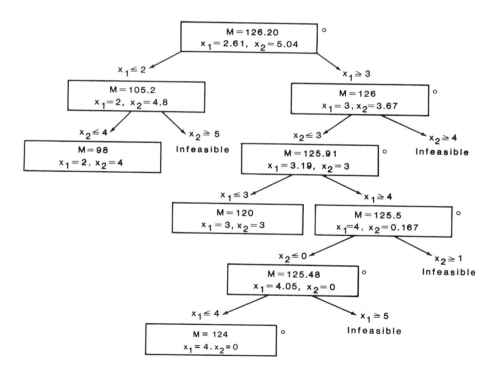

Figure 11–3

In example 5 the L. P. solution is $x_1 = 2.61$, $x_2 = 5.04$ with objective value $M = 126.20$. Variable x_1 is chosen for branching, producing two subcases— either $x_1 \leq 2$ or $x_1 \geq 3$. These two constraints divide the feasible region into two subregions with a missing strip between $x_1 = 2$ and $x_1 = 3$ that contains no grid points. The objective function may be maximized over each of the two subregions by linear programming. The results are shown in Figure 11–3 at the next tree level. Each solution is called a *node*. If the node is not integral then it has two subsequent branches. The node at $M = 105.2$ is temporarily set aside as $M = 126$ is more promising. These objective values represent upper bounds on the objective function for all subsequent nodes.

The two branches from the second node are $x_2 \leq 3$, $x_2 \geq 4$. As the diagram shows, the branch $x_2 \geq 4$ along with $x_1 \geq 3$ gives a region containing no points inside of the feasible region. The corresponding L. P. is infeasible. The remaining node with $M = 125.91$ produces two branches, $x_1 \leq 3$ and $x_1 \geq 4$. The first branch results in an integer solution at grid point (3, 3) with $M = 120$. The previous node at $M = 105.2$ is now fathomed, because 120 is a lower bound on M for integer solutions. The second branch at this tree level has a greater value for M than 120 so the tree must continue. Branches $x_2 \leq 0$ and $x_2 \geq 1$ give only one new node since the second branch is infeasible. The new node at $M = 125.48$ produces two final branches, $x_1 \leq 4$, $x_1 \geq 5$. Again the second branch is infeasible, but the first branch gives an integral solution at grid point (4, 0) with $M = 124$. All nodes are now terminal, so that 124 is the optimal objective of the integer program.

If another level were carried out from the node with $M = 105.2$, an integer solution would be found at point (2, 4). Its objective, $M = 98$, is necessarily less than or equal to 105. As soon as an integer solution is found with objective value M_1, all nodes with objective values $M < M_1$ are fathomed. •

The new constraint corresponding to a branch at any node may be added into the optimal tableau at that node in the same way that a Gomory constraint was added. The optimal tableau of example 5 with slack variables x_3 and x_4 is next.

	4	3	
2	.01709	.1795	5.0427
1	.04274	−.05128	2.6068
	1.4786	.02564	126.20

The second row was chosen for branching. Its corresponding equation is:

$$x_1 + .04274x_4 - .05128x_3 = 2.6068 \tag{1}$$

The two branches on x_1 may be written as equations with additional slack variables x_5, x_6.

$$x_1 + x_5 = 2 \tag{2}$$

$$x_1 - x_6 = 3 \tag{3}$$

The new constraint to be added to the optimal tableau at this node is found by subtracting (3) from (1) or by subtracting (1) from (2). The results are:

$$x_6 + .04274x_4 - .05128x_3 = - .3932 \tag{4}$$

$$x_5 - .04274x_4 + .05128x_3 = - .6068 \tag{5}$$

After adding in either (4) or (5) as an additional row, the tableau is infeasible and ready for a dual pivot. The change in the objective value due to this dual pivot is called a *penalty*, and it may be computed ahead of time. The *up penalty*, U, resulting from a dual pivot in the row corresponding to the upper branch (4), is:

$$U = \frac{.02564}{-.05128}(- .3932) = .20$$

If (4) is added as the ith row and q is the pivotal column found by the Dual Simplex Algorithm, then the formula is:

$$U = (Y_{mq} / Y_{iq}) Y_{in} \tag{6}$$

The *down penalty*, D, may be found in the same way by using the lower branch (5):

$$D = \frac{1.4786}{-.04274}(-.6068) = 21.00$$

The objective function will decrease by at least as much as the corresponding penalty for each branching. It might decrease more if further dual simplex pivots are necessary to gain primal feasibility. Thus, subtracting the penalty from M gives an upper bound on the objective for each possible branching. If this upper bound is less than some known lower bound, M_1 for integer solutions, the corresponding branch and all of its subsequent nodes are fathomed. In example 5:

$$M - U = 126 \quad \text{and}$$

$$M - D = 105.2.$$

The upper branch was picked for further branching, because it had the least penalty. After an integer solution was found with $M_1 = 120$, the lower branch originally skipped is fathomed and need not be considered.

Another useful penalty may be computed at each branching from the Gomory fractional cutting plane constraint. Recall that the Gomory constraint is made up from the fractional parts that must be subtracted from each coefficient of the generating row to reduce that coefficient to the next lower integer. Since the constraint is type II, each coefficient must be multiplied by minus one to put its slack variable into the basis.

In the example, the second row and its Gomory constraint are:

$$x_1 + .04274x_4 - .05128x_3 = 2.6068 \qquad\qquad (1)$$

$$s_1 - .04274x_4 - .94872x_3 = -.6068 \qquad\qquad (7)$$

where s_1 is the new slack variable. The penalty for adding (7) to the optimal tableau is the decrease in the objective value under one dual pivot in that row. Call it G, then:

$$G = \frac{.02564}{-.94872}(-.6068) = .0164$$

found by the same formula used in (6). An upper bound on the objective value is $M - G$. In this case min(U, D) $> G$, so that no smaller upper bound is found by using G.

A penalty that is available and does not depend on the generating row is the minimum of the shadow prices in the objective row. After pivoting in the new row some nonbasic variable must become positive, and if integer it is at least one. The formula for the change in the objective value,

$$\overline{Y}_{mn} = Y_{mn} - Y_{mq} (Y_{pn}/Y_{pq}),$$

shows that this change must be at least as much as the shadow price Y_{mq} because the other factor is at least one. Let S represent the minimum shadow price at any node. Then the three penalties may be combined as:

$$P = \max \{ S, G, \min (D, U) \}.$$

The objective value M at any node must decrease by at least as much as the single penalty P. Thus if $M - P \leq M_1$, a known lower bound, then the node at M is fathomed. Researcher J. A. Tomlin[5] reported that in his experience the bound P reduced the number of branches required to reach an optimum integer solution by as much as 50% compared to using the bound found from only up and down penalties.

---

[5] J. A. Tomlin, "An Improved Branch and Bound Method for Integer Programming," Operations Research, 19, (1971), 1073.

11.8   A BRANCH AND BOUND ALGORITHM

An important step in any branch and bound algorithm is the choice of a variable for branching at a node. Any variable that is noninteger may be used. One consideration is to try to pick a variable for which there is only one possible branch, because the other branch is infeasible. This situation occurs if one of the penalties is infinite. A variable is called *monotone increasing* as long as its down penalty is infinite and *monotone decreasing* as long as its up penalty is infinite. Such a choice holds down the tree size and eliminates "backtracking." A recommended choice for branching that makes use of monotone variables when present is:

1. Branch on the basic variable in row k where P_k = max$_i$ max (D_i, U_i), $i = 1, \ldots, m-1$.
2. If $P_k = D_k$ take the upper branch and if $P_k = U_k$ take the lower branch for x in the kth row.

Another possibility that the author has used effectively is based on the fractional parts of noninteger basic variables.

I. Branch on the basic variable in row k where F_k is the largest fractional part for a basic variable that satisfies

$$.1 \leq F_k \leq .9.$$

II. If no F_k exists in this restricted range, then pick F_k to be the largest fractional part available.

III. Compute the combined penalty P from section 11.7 for the row k picked by step I or II. If $M - P < M_1$, this node is fathomed. Otherwise choose the branch on x in row k corresponding to min (D_k, U_k).

A method must be devised to keep track of the live nodes during a tree search. R. J. Dakin suggested maintaining a "list" of the branches to be used, that is, the branches that lead to feasible solutions that are not fathomed by penalties.

When a terminal node is reached, the corresponding branch is dropped from the list. The remaining branches in the list are used to set up new L. P.s for which the process continues. Ultimately, when the "list" is empty, the tree search is complete and the best integer solution found is optimal.

ALGORITHM FOR BRANCH AND BOUND

1. Relax the integer requirement and solve the given problem as a L. P. If the solution is integer it is optimal. Otherwise, if it is nonintegral, go to step 2.

2. Establish a list for branches, initially empty. Estimate a lower bound, M_1, on the objective function for integer solutions.

3. Choose a branching variable by one of the methods discussed. Let k denote the number of the corresponding row.

4. For row k, compute the penalty

$$P = \max \{ S, G, \min (D, U) \}.$$

5. If the current objective value, M, satisfies $M - P < M_1$, go to step 11. Otherwise go to step 6 or 7.

6. If $\min (D_k, U_k) = D_k$, add the lower branch on x to the list and add its corresponding constraint to the current optimal tableau. If $M - U_k < M_1$, put a marker on the branch added to the list. The marker means that the upper branch for row k is fathomed. Go to step 8.

7. If $\min (D_k, U_k) = U_k$, add the upper branch on x to the list and add its corresponding constraint to the current optimal tableau. If $M - D_k < M_1$, put a marker on the branch added to the list as the lower branch is fathomed. Go to step 8.

8. Solve the current L. P.

9. If the new solution is infeasible, go to step 11, otherwise go to step 10.

10. If the new solution is feasible but not integer, go to step 3. If the feasible solution is integer and its objective is greater than M_1, replace M_1 with the new objective value. Go to step 11.

11. Is the list empty? If yes, the best integer solution found is optimal. Terminate. If no, go to step 12.

12. If the last entry in the list is marked, drop this constraint from the list and return to step 11. If the last entry is not marked, then replace the last entry with its alternate branch and mark this alternate branch. Add the entire list to the original problem constraints and go to step 8.

As an illustration of the branch and bound algorithm, consider again the bomb problem of Chapter 4 that was solved by Gomory's algorithms in sections 11.4 and 11.5. In step 1 start with the initial tableau as given in example 3 of section 11.5. The optimal tableau for the L. P. is

	6	2	5	
4	0	0	1	0
3	.3333	−1.444	−.2222	111.11
1	−.6667	3.222	.1111	444.44
7	−.6667	4.555	.7778	2111.1
	300.	1366.7	333.3	−966667

In step 2 estimate a lower bound on the objective function, $M_1 = -1,000,000$. Choose variable x_1 in the third row for branching. The penalty $P = \max \{ 300, 400, 188.5 \} = 400 = G$. Since the penalty does not eliminate this node, go to step 6 and find:

$$\min (D_3, U_3) = \min (188.5, 250) = 188.5 = D_3$$

Thus, the lower branch on x_1, namely $x_1 \leq 444$, is added to the list but not marked. The upper branch is still alive. The following constraint is added to the optimal tableau:

$$.6667 \, x_6 - 3.222 \, x_2 - .1111 \, x_5 \leq - .4444$$

Reoptimization leads to the next solution, called node (1). In what follows only a summary at each node will be given along with the "list" and a marker. The marker is 1 if the alternate branch is fathomed and 0 otherwise.

Node	List	Mark
(1) $M = 966,855$	$x_1 \leq 444$	0
$P = 652.5 = U$	$x_3 \geq 112$	0
$\min (5244.8, 652.5) = 652.5 = U$		
(2) $M = -967, 508$	$x_1 \leq 444$	0
$P = 492.3 = G = D$	$x_3 \geq 112$	0
$\min (492.3, 525.6) = 492.3 = D$	$x_2 \geq 1$	1

Branch $x_2 \leq 0$ is infeasible so is replaced by its alternate, $x_2 \geq 1$, and marked according to step 12.

(3) $M = -968,033$	$x_1 \leq 444$	0
$P = 420.5 = U$	$x_3 \geq 112$	0
$\min (500, 420.5) = 420.5 = U$	$x_2 \geq 1$	1
	$x_3 \geq 113$	0
(4) $M = -968,454$	$x_1 \leq 444$	0
$P = 246.1 = G = D$	$x_3 \geq 112$	0
$\min (246.1, 946.1) = 246.1 = D$	$x_2 \geq 1$	1
The down branch, $x_2 \leq 1$,	$x_3 \geq 113$	0
is infeasible, so it is	$x_2 \geq 2$	1
replaced with its alternate and marked.		
(5) $M = -969,400$		

Integer solution: $x_1 = 438$
$x_2 = 2$
$x_3 = 114$

According to step 10, replace the value of M_1 with $M_1 = -969,400$. The list is nonempty.

Following step 12, the last entry $x_2 \geq 2$ is dropped since it is marked. The next to last entry, $x_3 \geq 113$, is replaced with $x_3 \leq 112$, and marked.

Node		List	Mark
(6)	$M = -968{,}533$	$x_1 \leq 444$	0
	$P = 533.3 = S$	$x_3 \geq 112$	0
	min (2666.7, 300) = 300 = U	$x_2 \geq 1$	1
		$x_3 \leq 112$	1
		$x_1 \geq 443$	1

At this node the upper branch, $x_1 \geq 443$, is picked and marked because $M - D < M_1$ fathoms the lower branch. However, the current solution is infeasible, so the last three branches in the list are all dropped by steps 11 and 12. Branch $x_3 \geq 112$ is replaced with $x_3 \leq 111$ and marked.

(7)	$M = -972{,}100$	$x_1 \leq 444$	0
	Integer solution: $x_1 = 444$	$x_3 \leq 111$	1
	$\quad\quad\quad\quad\quad x_2 = 2$		
	$\quad\quad\quad\quad\quad x_3 = 111$		

This integer solution is eliminated because $M < M_1 = -969{,}400$. Drop the last marked entry in the list and replace the only branch left with $x_1 \geq 445$. Mark it.

(8)	$M = -966{,}917$	$x_1 \geq 445$	1
	$P = 383.3 = G = U = S$	$x_3 \leq 110$	1
	min (750,383.3) = 383.3 = U		

The upper branch is infeasible, so the lower branch, $x_3 \leq 110$, is added and marked.

(9)	$M = -967{,}667$	$x_1 \geq 445$	1
	$P = 533.3 = S = G$	$x_3 \leq 110$	1
	min (5333,150) = 150 = U	$x_1 \geq 447$	1

The upper branch is marked because $M - D < M_1$ fathoms the lower branch.

Node		List	Mark
(10)	$M = -967{,}817$	$x_1 \geq 445$	1
	$P = 383.3 = S = G = U$	$x_3 \leq 110$	1
	min (750,383.3) = 383.3 = U	$x_1 \geq 447$	1
	The upper branch, $x_3 \geq 110$,	$x_3 \leq 109$	1
	is infeasible so $x_3 \leq 109$		
	is added to the list and marked.		
(11)	$M = -968{,}567$	$x_1 \geq 445$	1
	$P = 533.3 = S = G$	$x_3 \leq 110$	1
	min (5333,150) = 150 = U	$x_1 \geq 447$	1
	The upper branch is	$x_3 \leq 109$	1
	added to the list and	$x_1 \geq 449$	1
	marked because $M - D < M_1$		
	fathoms the lower branch.		

(12) $M = -968,717$ $x_1 \geq 445$ 1
      $P = 383.3 = S = G = U$ $x_3 \leq 110$ 1
      min (750,383.3) = 383.3 = U $x_1 \geq 447$ 1
      The upper branch, $x_3 \geq 109$, $x_3 \leq 109$ 1
      is added and again it is $x_1 \geq 449$ 1
      marked because $M - D < M_1$. $x_3 \geq 109$ 1

The current solution is infeasible. Since all of the entries in the list are marked, the list drops to zero entries and the algorithm terminates. The integer solution found at node (5) is the optimal integer solution and agrees with the previous solutions. The typical situation in a tree search is that an integer solution is found fairly early in the search, but many more nodes must be examined to complete the proof of optimality. ●

Many mathematical programming problems require only some of the variables to be integer. Both the fractional cutting plane and the branch and bound algorithms presented in this chapter may be terminated when the desired variables are integer. Branches or cuts should be chosen only among the variables that are required to be integer.

All of these techniques of linear and integer programming are an outgrowth of the simple and elegant Simplex Method of George Dantzig. This has been one of the most fruitful concepts in modern mathematics. It has allowed us to use the largest and fastest computers to solve monumental and everyday practical problems.

Problems

Solve by Branch and Bound

11–38. Show the tree of branches and also the graph for solving the I.P.

$x_i \geq 0$ and integer for $i = 1,2$

$3x_1 - x_2 \leq 3$

$-x_1 + 3x_2 \leq 3$

$x_1 + x_2 = M$, a maximum.

11–39. Find nonnegative integers x_1 and x_2 that satisfy

$x_1 + 8x_2 \leq 32$

$2x_1 + x_2 \leq 14$

$3x_1 + 20x_2 = M$, a maximum.

Does the integer solution agree with a rounded general solution?

11-40. Find vector **X** and the maximum, M, such that:

$$x_i \geq 0 \text{ and integer for } i = 1,2,3$$
$$x_1 - 2x_2 + 3x_3 \leq 24$$
$$3x_1 + 7x_2 + 4x_3 \leq 86$$
$$x_1 + 3x_2 + 7x_3 \leq 45$$
$$x_1 \qquad + 4x_3 = M.$$

Find an alternate integer optimum.

11-41. Find nonnegative integers x_1 and x_2 such that:

$$6x_1 + 10x_2 \leq 69$$
$$3x_1 + 10x_2 \leq 60$$
$$6x_1 - 15x_2 \leq 24$$
$$5x_1 + 2x_2 = M, \text{ a maximum.}$$

11-42. Find a nonnegative integer solution and the maximum, M, for

$$22.12x_1 + 34.31x_2 \leq 105.26$$
$$1.01x_2 \leq 1.87$$
$$x_1 + 3x_2 = M.$$

Do the entries in the initial tableau need to be integers to get an integer solution?

11-43. Find two nonnegative integer solutions such that

$$12000x_1 + 7000x_2 \leq 55000$$
$$2000x_1 + 1000x_2 \geq 7500$$
$$250x_1 + 100x_2 \leq 900$$
$$20000x_1 + 10000x_2 = M, \text{ a maximum.}$$

11-44. Find as many integer solutions as possible by the branch and bound method for

$$.05x_1 + x_2 \geq 1.8$$
$$-.9x_1 + x_2 \leq 2.2$$
$$x_1 + .65x_2 \leq 5.4$$
$$x_2 = m, \text{ a minimum.}$$

Solve the problem graphically and notice there are 5 integer solutions, but only two of them are found by branch and bound.

11-45. Find the minimum and nonnegative integer solution for

$$
\begin{aligned}
-x_1 \quad\quad + 3x_3 \quad\quad &\geq 1 \\
-15x_1 - 3x_2 + 7x_3 + \quad x_4 &= 21 \\
-3x_1 + \quad x_2 + 2x_3 \quad\quad &\leq 3 \\
9x_1 + 4x_2 + \quad x_3 + \quad x_4 &= m.
\end{aligned}
$$

11-46. A knapsack problem requires the packing of a knapsack with items of a given value and weight. The problem is to decide which of the items to include to maximize the value to be carried in the knapsack while satisfying the upper bound on the total weight. Solve the following case where there are five times to choose from whose weight and value are given in the table:

	ITEM				
	1	2	3	4	5
pounds	36	24	30	32	26
dollars	54	18	60	32	33

The maximum weight allowed is 91 pounds. The 5 unknowns have only two values, zero if the item is left out and one if the item is packed in the knapsack.

Given here are 14 knapsack problems that differ only by the upper bound on weight to be carried in the knapsack. Solve each problem and note the cases that have alternate optima. As the upper bound on weight increases see that the number of items to be carried increases up to all ten items. The form for each problem is:

Maximize:

$$M = 20x_1 + 18x_2 + 17x_3 + 15x_4 + 15x_5 + 10x_6 + 5x_7 + 3x_8 + x_9 + x_{10}$$

such that:

$$30x_1 + 25x_2 + 20x_3 + 18x_4 + 17x_5 + 11x_6 + 5x_7 + 2x_8 + x_9 + x_{10} \leq b$$

where $x_i = 1, 0$ for $i = 1, 2, \ldots, 10$.

The following table gives the constants b for each of the problems:

PROBLEM	b
11–47	35
11–48	45
11–49	55
11–50	60
11–51	65
11–52	70
11–53	75
11–54	80
11–55	85
11–56	90
11–57	100
11–58	110
11–59	120
11–60	130

Problems 11–49 through 11–57 are the same as the nine allocations problems tested by C.A. Trauth and R.E. Woolsey (1969)[7].

Fixed Charge Problems
The following ten fixed charge problems were originated by J. Haldi (1964)[8] and were included in the C. A. Trauth and R. E. Woolsey study (1969). A fixed charge problem is characterized by a fixed cost such as a "start up" cost or investment cost that is incurred only if the corresponding project is chosen. Such a problem may be formulated as an Integer Program by adding in binary variables x_i, where $x_i = 1$ if its designated project is carried out and $x_i = 0$ if its project is dropped. In the form given for these problems, the fixed charges are the coefficients of x_1 and x_2 in the constraints with upper bounds B_i. In some of Haldi's problems the variables x_1 and x_2 are not restricted to be binary but are still integer. If they are greater than one then the corresponding fixed charges would be incurred more than once for an activity that is repeated. The constraints give upper bounds on the resources available and the objective is to maximize some output.

Problems 11–61 to 11–64 follow the format:
Maximize $M = x_3 + x_4 + x_5$

such that:

---

[7] Trauth, C. A., and Woolsey, R. E., Integer Linear Programming: A Study in Computational Efficiency, *Management Science* 15, 9 (1969), 481–493.
[8] Haldi, J. 25 Integer Programming Test Problems, Working Paper No. 43, Graduate School of Business, Stanford University (Dec. 1964).

$$2x_1 + \quad 3x_2 + \quad x_3 + 2x_4 + 2x_5 \le B_1$$

$$3x_1 + \quad 2x_2 + 2x_3 + \quad x_4 + 2x_5 \le B_2$$

$$-R_1 x_1 \qquad\qquad + \quad x_3 \qquad\qquad\quad \le \ 0$$

$$-R_2 x_2 \qquad\qquad + \quad x_4 \quad\quad\ \ \le \ 0$$

where x_1 is a nonnegative integer for $i = 1,2,3,4,5$. The following table gives the constants B_i and R_i for each problem.

Problem	R_1	R_2	B_1	B_2
11–61	6	7	18	15
11–62	9	7	18	17
11–63	9	9	21	21
11–64	6	8	19	15

Problems 11–65 and 11–66 are:
  Maximize $M = x_3 + x_4 + x_5$

such that:

$$20x_1 + \quad 30x_2 + \quad x_3 + 2x_4 + 2x_5 \le B_1$$

$$30x_1 + \quad 20x_2 + 2x_3 + \quad x_4 + 2x_5 \le B_2$$

$$-R_1 x_1 \qquad\qquad + \quad x_3 \qquad\qquad\quad \le \ 0$$

$$-R_2 x_2 \qquad\qquad + \quad x_4 \quad\quad\ \ \le \ 0$$

$$x_1 \qquad\qquad\qquad\qquad\qquad \le \ 1$$

$$x_2 \qquad\qquad\qquad\qquad\qquad \le \ 1$$

where x_i is a nonnegative integer for $i = 1,2,3,4,5$. The constants B_i and R_i are given as follows:

Problem	R_1	R_2	B_1	B_2
11–65	60	75	180	150
11–66	90	90	210	210

Problem 11–67 is the same problem as 11–65 except the two binary constraints on x_1 and x_2 are dropped. Problem 11–68 is the same as problem 11–66 except that the final two constraints are dropped. Variables x_1 and x_2 appear free to be larger than one, although the answers show that this doesn't happen.

Problems 11–67 and 11–68 are extremely difficult to solve by cutting planes. When you use branch and bound, do the solutions or times of execution differ from problems 11–65 and 11–66?

11–69. Problem 11–69 is as follows:
Maximize $M = x_4 + x_5 + x_6$

such that:

$$
\begin{aligned}
2x_1 + \ 2x_2 \ \ \ \ \ \ \ \ \ + \ x_4 + x_5 \ \ \ \ \ \ \ &\leq 10 \\
2x_1 \ \ \ \ \ \ \ + \ 2x_3 + \ x_4 \ \ \ \ \ \ + \ x_6 &\leq 10 \\
2x_2 + \ 2x_3 \ \ \ \ \ \ \ + \ x_5 \ + \ x_6 &\leq 10 \\
-8x_1 \ \ \ \ \ \ \ \ \ \ \ \ \ \ \ + \ x_4 \ \ \ \ \ \ \ \ \ \ \ \ &\leq 0 \\
-\ 8x_2 \ \ \ \ \ \ \ \ \ \ \ \ \ \ + \ x_5 \ \ \ \ \ \ &\leq 0 \\
-\ 8x_3 \ \ \ \ \ \ \ \ \ \ \ + \ x_6 &\leq 0
\end{aligned}
$$

where all x_i, $i = 1, 2, \ldots, 6$, are nonnegative integers.

11–70. Problem 11–70 is given in the table below:

	x_1	x_2	x_3	x_4	x_5	x_6	x_7	x_8	x_9	x_{10}	x_{11}	x_{12}		
$M =$							1	1	1	1	1	1		
	9	7	16	8	24	5	3	7	8	4	6	5	$\leq$	110
	12	6	6	2	20	8	4	6	3	1	5	8	$\leq$	95
	15	5	12	4	4	5	5	5	6	2	1	5	$\leq$	80
	18	4	4	18	28	1	6	4	2	9	7	1	$\leq$	100
	−12						1						$\leq$	0
		−15						1					$\leq$	0
			−12						1				$\leq$	0
				−10						1			$\leq$	0
					−11						1		$\leq$	0
						−11						1	$\leq$	0

where M is maximized and all twelve variables are nonnegative integers.

The next four problems, called the IBM® problems because the IBM Corporation supplied them, were also used by Haldi in his study and were included in the Trauth and Woolsey study. Originally, there were nine of these problems, which are considered classics for testing integer programming codes. The last problem given, called IBM10, is a slight variation of IBM9.

11–71. *IBM1* Solve:

	x_1	x_2	x_3	x_4	x_5	x_6	x_7		b_1
$m =$	1	1	1	1	1	1	1		
	1			1	1		1	$\geq$	5
		1		1		1	1	$\geq$	5
			1		1	1	1	$\geq$	5
	1	1			1	1		$\geq$	4
	1		1	1		1		$\geq$	4
		1	1	1	1			$\geq$	4
	1	1	1				1	$\geq$	3

where m is a minimum and all x_i, $i = 1,2,\ldots,7$, are nonnegative integers.

11–72. *IBM2* Solve:

	x_1	x_2	x_3	x_4	x_5	x_6	x_7	b_2
$m =$	1	1	1	1	1	1	1	
	1			1	1		1	$\geq$ 4
		1		1		1	1	$\geq$ 4
			1		1	1	1	$\geq$ 4
	1	1			1	1		$\geq$ 3
	1		1	1		1		$\geq$ 3
		1	1	1	1			$\geq$ 3
	1	1	1				1	$\geq$ 2

where m is a minimum and all x_i, $i = 1,2,\ldots,7$, are nonnegative integers.

11–73. *IBM3*

IBM3 is given as follows:

Minimize $13x_1 + 15x_2 + 14x_3 + 11x_4$

such that

$$4x_1 + 5x_2 + 3x_3 + 6x_4 \geq 96$$
$$20x_1 + 21x_2 + 17x_3 + 12x_4 \geq 200$$
$$11x_1 + 12x_2 + 12x_3 + 7x_4 \geq 101$$

where x_i, $i = 1,2,3,4$, are nonnegative integers.

11–74. *IBM9*

Problem IBM9 has a 50 by 15 coefficient matrix, where each column has 8 ones and 42 zeros. Table I gives, under each column heading, the eight rows that contain the ones. All 50 constants, b, on the right-hand side of the inequalities, are one. Rows 1 through 35 have inequalities, $\geq$, while rows 36 through 50 have inequalities, $\leq$. The objective is to minimize the sum of the fifteen variables.

TABLE I
Problem IBM9

Column	1	2	3	4	5	6	7	8	9	10	11	12	13	14	15
min $m =$	1	1	1	1	1	1	1	1	1	1	1	1	1	1	1
Enter "1"	3	4	1	1	2	1	2	3	4	5	11	12	13	14	15
on the	4	5	5	2	3	6	7	8	9	10	16	17	18	19	20
indi-	7	6	7	8	6	13	14	11	11	12	23	24	21	21	22
cated	10	8	9	10	9	14	15	15	12	13	24	25	25	22	23
rows	21	22	23	24	25	17	16	17	18	16	27	26	27	28	26
	26	27	28	29	30	20	18	19	20	19	30	28	29	30	29
	31	32	33	34	35	31	32	33	34	35	31	32	33	34	35
	36	37	38	39	40	41	42	43	44	45	46	47	48	49	50

IBM9 has the form of a set covering problem. For a given index set I, a "cover" is a collection of subsets of I such that each integer in I lies in at least one subset of the collection. Let $I = \{1, 2, \ldots, m\}$ and $\{P_1, P_2, \ldots, P_n\}$ be a collection of subsets where each P_j is contained in I and j is in set $J = \{1, 2, \ldots, n\}$. The cover is a subset of J such that the union of P_j, for j in that subset, equals I. If each j in J has an associated cost, then the set covering problem is to find the cover of minimum cost. As an integer program, the problem is:

$$\min \sum_{j=1}^{n} c_j x_j$$

$$\sum_{j=1}^{n} a_{ij} x_j \geq 1, \ i = 1, 2, \ldots, m$$

$$x_j = 0 \text{ or } 1, j = 1, 2, \ldots, n$$

where $\qquad x_j = 1$ if j is in the cover

$\qquad\qquad\qquad = 0$ otherwise

and $\qquad a_{ij} = 1$ if i is in P_j

$\qquad\qquad\qquad = 0$ otherwise.

In IBM9, the index sets are determined by $m = 35$ and $n = 15$. The final fifteen constraints merely say that the fifteen variables, x_j, are binary. All of the cost coefficients are 1 so that the objective is to find the minimum cover.

11–75. Trauth and Woolsey (1969) have a misprint in their coefficient matrix for IBM9. In their tenth column, the first 1 should be in row 5 instead of row 15. This misprint gives rise to a totally different problem, which has fewer alternate optima than the original problem 9 and will be called IBM10. The new problem, IBM10, has a minimum value of 8.

The really tough problems have massive dual degeneracy, that is, the relative costs or shadow prices are mostly zero. This occurs in branch and bound iterations when the problem has a large number of alternate optima. If a problem can be formulated so as to reduce the number of optima, then the branch and bound performance is enhanced. Such a change is noted in going from IBM9 to IBM10 although the minimum value is reduced from 9 to 8. The branch and bound technique is especially affected by "tying" nodes where many nodes have the same objective value. In column 10 of Table I, change 5 to 15, i.e., place the 1 in row 15 instead of row 5. Now solve the new problem, IBM10, and see the mentioned effects.

Appendix

SUBROUTINES FOR PRIMAL-DUAL ALGORITHM

In Chapter 5 a set of subroutines was given for the automatic pivoting of a linear programming tableau where artificial variables, if present, were removed from the basis. This appendix contains an alternate set of subroutines for the automatic pivoting of an initial tableau. The technique programmed here avoids artificial variables by the methods of Chapter 8. The following subroutines are to be used with a program that follows the Primal-Dual Algorithm for Mixed Systems given in Section 8.2.

The version of BASIC used in the following subroutines is called BASIC8 in the development of the language at Dartmouth University. BASIC8 differs from the earlier versions of BASIC in the CALL statements and SUB statements. The remaining statements in these subroutines are compatible with other versions of BASIC. If you use a version of BASIC or FORTRAN that differs from those presented here, some minor adjustments may need to be made.

The initial tableau is to be set up according to the first four steps of the algorithm in Section 8.2. Thus, all of the constraints have been converted to type I inequalities ($\leq$), and the objective function is a function to be maximized. The resulting initial tableau may be infeasible or nonoptimal or both. All possible pivots will be considered by both the Simplex Method and the Dual Simplex Method.

To carry out the algorithm's fifth step, first call SUBROUTINE DROW.

FORTRAN

```
      SUBROUTINE DROW ( Y,M,N,IP,W )
      DIMENSION Y ( 30,30 ),W( 30 )
      IP=−1
      B=0.
      MM=M−1
      DO 935 I=1,MM
      IF( Y( I,N )+.1E−5 ) 915,935,935
  915 IF ( W( I ) ) 920,920,935
  920 IF( Y( I,N )−B ) 925,935,935
  925 B=Y( I,N )
      IP=I
  935 CONTINUE
      RETURN
      END
```

BASIC

```
1490    SUB DROW ( Y( , ), W( ), M, N, P )
1500    LET P=−1
1510    LET B=0
1520    FOR I=1 TO M−1
1530    IF ( Y ( I,N ) +1E−4 )>=0 THEN 1580
1540    IF W (I )>0 THEN 1580
1550    IF ( Y( I,N )−B )>=0 THEN 1580
1560    LET B=Y ( I,N )
1570    LET P=I
1580    NEXT I
1590    END SUB
```

The vector $\mathbf{W}$ is to be initialized by the main program at $\mathbf{W}(I)= 0$ for $I= 1$ to M. It will be used to skip the Ith row if that row has the most negative number in its last column but no possible pivot. Then the next most negative number in the last column will determine a possible pivotal row. The first IF statement in DROW is for the purpose of skipping a row with either zero or a positive number in its last column. Note the epsilon that has been added as a bow to the inaccuracies of the computer. If $Y(I,N)$ should be zero but turns up as a small negative number due to round-off error, then the small positive number will correct the problem of choosing that row. The second IF statement causes the Ith row to be skipped whenever $W(I)$ is positive. The remaining statements are similar to the corresponding statements in subroutine COL and cause the most negative number available, $Y(I, N)$, to pick the pivotal row. Of course, if there is no possible pivotal row, P or IP is returned to the main program at -1. If DROW returns a positive P or IP, then that value should be stored for future reference and subroutine DCOL called.

FORTRAN

```
        SUBROUTINE DCOL ( Y,M,N,IQ,IP,DS )
        DIMENSION Y ( 30,30 )
        IQ=-1
        DS=1.E19
        NN=N-1
        DO780 J=1,NN
        IF( Y( IP,J )+.1E-5 )740,780,780
    740 IF( Y( M,J )+.1E-5 )780,750,750
    750 R=-Y( M,J )/Y( IP,J )
        IF( DS-R )780,780,760
    760 DS=R
        IQ=J
    780 CONTINUE
        RETURN
        END
```

BASIC

```
1380 SUB DCOL ( Y ( , ), M, N, P, Q, S ) )
1390 LET G=-1
1400 LET S=1E19
1410 FOR J=1 TO N-1
1420 IF ( Y ( P,J )+1E-4 )>=0 THEN 1480
1430 IF Y ( M,J )+1E-4 )<0 THEN 1480
1440 LET R=-Y( M,J )/Y( P,J )
1450 IF ( S-R )<=0 THEN 1480
1460 LET S=R
1470 LET Q=J
1480 NEXT J
1490 END SUB
```

This subroutine looks for a pivotal column by examining the $\bar{\theta}$ ratios between the last row and the pivotal row. The first IF statement will skip any zero or positive entry in the pivotal row. Again note the small positive number that has been added to take care of round-off error. For a negative entry in the pivotal row, the second IF statement will skip that column if there is a negative number in the bottom row. Thus, the $\bar{\theta}$ ratios will be negative or zero. Statement 750 or 1440 computes the positive value of the $\bar{\theta}$ ratios. The next two statements are similar to the corresponding statements in subroutine ROW. They pick the $\bar{\theta}$ ratio of smallest numerical value. Then Q or IQ is set equal to the number of that column. If there is no possible pivotal column, then Q or $IQ = -1$ is returned. In this case, the main program should set $\mathbf{W}(IP) = 1$ or $\mathbf{W}(P) = 1$ and call DROW again. Thus row IP or P will be skipped as we try to find another pivotal row.

This search is continued until both IP and IQ or both P and Q return positive or else there is no dual pivot. If no pivot has been found, proceed directly to step six of our algorithm. When a pivot has been determined, store its position in memory. Then compute the decrease in objective value by the formula

$$D_q = ABS(Y(IP,N) \cdot DS).$$

The DS is the smallest numerical value of the $\bar{\theta}$ ratios found by DCOL in determining the pivotal column. We are now ready for step six of the Primal-Dual Algorithm.

In order to find a pivot by the Simplex Method, we use two subroutines that are only slightly different from COL and ROW found in Chapter 5. The first is COL2 that will search for a pivotal column.

FORTRAN

```
      SUBROUTINE COL2 ( Y,M,N,IQ,V )
      DIMENSION Y ( 30,30 ),V( 30 )
      IQ=-1
      B=0.
      NN=N-1
      DO 30 J=1,NN
      IF( V( J ) )10,10,30
   10 IF( Y( M,J )+.1E-5 )20,30,30
   20 IF( Y( M,J )-B )25,30,30
   25 B=Y( M,J )
      IQ=J
   30 CONTINUE
      RETURN
      END
```

BASIC

```
1000  SUB COL2 ( Y( , ), V( ), M, N, Q )
1010  LET Q=−1
1020  LET B=0
1030  FOR J=1 TO N−1
1040  IF ( Y ( M,J )+1E−4 )>=0 THEN 1090
1050  IF V ( J )>0 THEN 1090
1060  IF ( Y( M,J )−B )>=0 THEN 1090
1070  LET B=Y( M,J )
1080  LET Q=J
1090  NEXT J
1100  END SUB
```

The vector $V(J)$ should initially be set equal to zero for $J = 1$ to N. It is used to skip column J when no pivot is available in column J. Its role is analogous to that of $W (I)$ in DROW. Column J will be skipped whenever $V(J)$ is positive. In the first IF statement, an epsilon has been added to take care of round-off error. The same correction could be added to COL in Chapter 5. The remainder of COL2 is similar to COL.

If $Q = −1$ or $IQ = −1$ is returned by COL2, then the pivot chosen by step five above will be used or else the pivoting is completed. If Q or IQ returns positive, it will be stored and subroutine ROW2 called.

FORTRAN

```
      SUBROUTINE ROW2 ( Y,M,N,IP,IQ,S )
      DIMENSION Y( 30,30 )
      IP=−1
      S=1.E19
      MM=M−1
      DO 80 I=1,MM
      IF( Y( I,IQ )−.1E−5 )80,80,5
    5 IF( Y( I,N )+.1E−5 )80,10,10
   10 R=Y( I,N )/Y( I,IQ )
      IF( S−R )80,80,20
   20 S=R
      IP=I
   80 CONTINUE
      RETURN
      END
```

BASIC

```
1100  SUB ROW2 ( Y ( , ), M,N,P,Q,S )
1110  LET P=−1
1120  LET S=1E19
1130  FOR I=1 TO M−1
1140  IF ( Y ( I,Q )−1E−4 )<=0 THEN 1200
1150  IF Y ( I,N )+1E−4 )<0 THEN 1200
1160  LET R=Y( I,N )/Y( I,Q )
1170  IF ( S−R )<=0 THEN 1200
1180  LET S=R
1190  LET P=I
1200  NEXT I
1210  END SUB
```

The first IF statement will skip any negative or zero entries in the pivotal column. Notice that the epsilon value has been subtracted. This will eliminate any rows with zero values that turn out to be positive values due to round-off error. The same correction could be made to ROW in Chapter 5. For the positive entries in the pivotal column, the second IF statement will skip those rows with a negative number in the last column. Again the epsilon correction has been added so that we do not skip any negative numbers that should be zero. Thus, the θ ratios considered will be positive or zero. The remaining statements that pick the minimum θ ratio and set the pivotal row are similar to those of ROW.

If ROW2 returns a negative for the pivotal row, then set $V(Q) = 1$ or $V(IQ) = 1$ and call COL2 again to look for the next possible pivotal column. Since any previously examined columns will be skipped in COL2, only new possible pivotal columns will be considered before returning to ROW2. Continue this procedure until either no pivot is found or else both the row and column choices are positive. In case they are both positive, compute the increase in objective value by the formula

$$I_p = \text{ABS}(Y(M,IQ)^\ast S).$$

S is the minimum value of the considered θ ratios. If no pivot were found by the Dual Simplex Method, D_q could have been set equal to -1. Now compare I_p with D_q. Pick the larger of the two and set the pivot position (P, Q) or (IP, IQ) equal to the values that were stored by the corresponding method. Using this pivot, a call to PIVCO, found in Chapter 5, will carry out a pivoting iteration at the chosen position.

After completing PIVCO, the vectors $W(I)$ and $V(J)$ must be reset to zero before searching for the next pivot position. As soon as no new pivot can be found, the final tableau has been reached and should be printed out for interpretation. The analysis of the final tableau is found in Section 8.2 following the Primal-Dual Algorithm.

Index